D0282767

Modern studies of the atmosphere, oceans, and earth and planetary systems demand a sound knowledge of basic chemical principles. This book provides a clear, concise review of these principles, particularly relating to the atmosphere. Students with little formal training in chemistry can work through the chapters and numerous exercises within this book before accessing the standard texts in the atmospheric and environmental sciences.

Basic Physical Chemistry for the Atmospheric Sciences covers the fundamental concepts of chemical equilibria, chemical thermodynamics, chemical kinetics, solution chemistry, acid and base chemistry, oxidation–reduction reactions, and photochemistry. Over 160 exercises are contained within the text, including 50 numerical problems solved in the text and 112 exercises for the reader to work on with hints and solutions provided in an appendix.

Basic Physical Chemistry for the Atmospheric Sciences is written by an internationally known atmospheric science researcher and successful textbook author. Whether used in an introductory class in chemistry or for self-instruction as a supplementary workbook this book will be an invaluable resource for a broad range of students.

BASIC PHYSICAL CHEMISTRY FOR THE ATMOSPHERIC SCIENCES

BASIC PHYSICAL CHEMISTRY FOR THE ATMOSPHERIC SCIENCES

PETER V. HOBBS

University of Washington

CAMBRIDGE
UNIVERSITY PRESS

Published by the Press Syndicate of the University of Cambridge
The Pitt Building, Trumpington Street, Cambridge CB2 1RP, England
40 West 20th Street, New York, NY 10011-4211, USA
10 Stamford Road, Oakleigh, Melbourne 3166, Australia

First published 1995

Printed in the United States of America

Library of Congress Cataloging-in-Publication Data
Hobbs, Peter Victor, 1936–
Basic physical chemistry for the atmospheric sciences / Peter
V. Hobbs.
p.   cm.
Includes index.
ISBN 0–521–47387–X. — ISBN 0–521–47933–9 (pbk.)
1. Atmospheric sciences.   2. Atmospheric chemistry.   3. Chemistry,
Physical and theoretical.   I. Title.
QC861.2.H63   1995
551.5'11—dc20   94–19216
CIP

A catalog record for this book is available from the British Library.

ISBN 0–521–47387–X hardback
0–521–47933–9 paperback

Contents

Preface

A short account of the origins of this book will explain its purpose. In the 1970s I coauthored (with John M. Wallace) a textbook for senior undergraduates and first-year graduate students entitled *Atmospheric Science: An Introductory Survey* (Academic Press, 1977). At the time that text was written it was not considered necessary to include a chapter on atmospheric chemistry. By the early 1990s, when we began to think about a second edition of *Atmospheric Science,* the importance of atmospheric chemistry was such that it was inconceivable that such a book would not include a substantial chapter on this subject.

In the intervening years I had introduced a section on atmospheric chemistry into the survey course taken by all first-year graduate students in the Atmospheric Sciences Department at the University of Washington. I quickly discovered, however, that many of the students either had no previous instruction in chemistry or had long since forgotten what little they had known. I therefore wrote an (unpublished) primer on physical chemistry for these students; the present book grew out of that primer.

Reviewed herein are some of the fundamental concepts associated with chemical equilibrium, chemical thermodynamics, chemical kinetics, aqueous solutions, acid–base chemistry, oxidation–reduction reactions and photochemistry, all of which are essential to an understanding of atmospheric chemistry. The approach is primarily from the macroscopic viewpoint, which provides the tools needed by the pragmatist. A deeper understanding requires extensive treatment of the electronic structure of matter and chemical bonding, topics that are beyond the scope of this introductory text. This book can be used for either self-instruction, or as the basis for a short introductory class

ix

on chemistry, prior to courses in which chemistry is applied to one of the geosciences. In addition to students (and I use this term in its broadest sense) of atmospheric sciences, I hope this book will be useful to others. It should be suitable, for example, as a precursor to undergraduate and graduate courses in which chemistry is applied to any of the geosciences and environmental sciences.

In keeping with the didactic approach of this book, and the view that any science is best learned by solving problems, I have provided solutions to 50 exercises in the text and posed 112 exercises for the student. Answers to all the quantitative problems, and hints and solutions to selected problems, are given in Appendix VII.

In preparing this book I benefited from the following texts, which are recommended to the reader. *Chemistry: An Experimental Science* edited by G. C. Pimentel (W. H. Freeman, 1963); this book gives a broad introduction to chemistry with emphasis on its experimental foundations. *Chemistry: The Central Science* by T. L. Brown and H. E. LeMay Jr. (Prentice-Hall Inc., 1981) and *General Chemistry: Principles and Modern Applications* by R. H. Petrucci (Macmillan Pub. Co., 1982), which provide more extensive accounts of most of the topics discussed in the present book as well as dealing with many other aspects of chemistry. Finally, for the student who wants to take the next step in chemistry beyond that presented here, *University Chemistry* by B. H. Mahan (Addison-Wesley, 1965) is highly recommended.

This book was started in 1984 when I was an Alexander von Humboldt Foundation Senior Scientist in Germany, and it was essentially finished in 1993 during a sabbatical at the Instituto FISBAT-CNR, Bologna, Italy. Thanks are due to both of these organizations for their generous support. It is also a pleasure to thank my colleagues Professors Dean Hegg and Conway Leovy, and many students, particularly John Herring and Cathy Cahill, who commented on various drafts of this book and made suggestions for its improvement. I am grateful to the National Sciences Foundation for supporting my own research in atmospheric chemistry over many years.

Any suggestions or corrections related to this book will be gratefully received.

Seattle
May 1994

1

Chemical equilibrium

One of the major goals of chemistry is to predict what will happen when various substances come into contact. Will a chemical reaction occur, or will the substances just exist side by side? One to way to approach this problem is through the concept of chemical equilibrium, which is the focus of this chapter.

1.1 Some introductory concepts

In a balanced equation for a chemical reaction, there are the same number of atoms of each element on the left side of the equation as there are on the right side. For example, the balanced equation for the chemical reaction representing photosynthesis is[1]

$$6CO_2(g) + 6H_2O(l) \rightarrow$$
$$C_6H_{12}O_6(s) + 6O_2(g) \tag{1.1a}$$

In a balanced chemical equation (which we will often call a *reaction*), the relative numbers of the molecules involved in the reaction are given by the numerical coefficients preceding the chemical symbol for the molecule. Thus, Reaction (1.1a) indicates that six molecules of carbon dioxide, $CO_2(g)$, react with six molecules of water, $H_2O(l)$, to form one molecule of glucose, $C_6H_{12}O_6(s)$, and six molecules of oxygen, $O_2(g)$

$$6 \text{ molecules of } CO_2(g) + 6 \text{ molecules of } H_2O(l) \rightarrow$$
$$1 \text{ molecule of } C_6H_{12}O_6(s) + 6 \text{ molecules of } O_2(g) \tag{1.1b}$$

Reaction (1.1a) does not necessarily mean that if six molecules of $CO_2(g)$ are mixed with six molecules of $H_2O(l)$ they will react completely and produce one molecule of $C_6H_{12}O_6(s)$ and six molecules of

$O_2(g)$. Some chemical reactions proceed very quickly, others very slowly; and some never reach completion. However, what Reaction (1.1a) does tell us is that at any given instant in time the ratio of the numbers of molecules of $CO_2(g)$, $H_2O(l)$, $C_6H_{12}O_6(s)$, and $O_2(g)$ that have reacted is 6:6:1:6.

The relative masses of the various atoms are represented by their *atomic weights* (dimensionless) referenced to carbon-12 (i.e., a carbon atom containing six protons and six neutrons), where carbon-12 is arbitrarily assigned an exact atomic weight of 12. Atomic weights are listed in Appendix III. Similarly, the relative masses of molecules are represented by their *molecular weights* (dimensionless), where the molecular weight is obtained by adding together the atomic weights of all the atoms in the molecule. For example, since the atomic weights of hydrogen and oxygen are 1.008 and 15.999, respectively, the molecular weight of water (H_2O) is $(2 \times 1.008) + 15.999 = 18.015$.

One *gram-molecular weight* (abbreviation *mole* or *mol*) of any compound is a mass of that compound equal to its molecular weight in grams. Thus, 1 mole of water is 18.015 g of water. One mole of any compound contains the same number of molecules as one mole of any other compound.[2] The number of molecules in 1 mole of any compound is 6.022×10^{23}, which is called *Avogadro's number* (N_A). Since the volume occupied by a gas depends on its temperature, pressure, and the number of molecules in the gas, at the same temperature and pressure 1 mole of the gas of any compound occupies the same volume as 1 mole of the gas of any other compound. At *standard temperature and pressure* (STP), which are defined as 0°C and 1 bar ($= 10^5$ Pa),[3] the volume occupied by 1 mole of any gas is about 22.4 L.

If we now multiply every term in the Relation (1.1b) by N_A we get

$$6N_A \text{ molecules of } CO_2(g) + 6N_A \text{ molecules of } H_2O(l) \rightarrow$$
$$1 N_A \text{ molecule of } C_6H_{12}O_6(s) + 6N_A \text{ molecules of } O_2(g)$$

or,

$$6 \text{ moles of } CO_2(g) + 6 \text{ moles of } H_2O(l) \rightarrow$$
$$1 \text{ mole of } C_6H_{12}O_6(s) + 6 \text{ moles of } O_2(g) \qquad (1.1c)$$

Relations (1.1b) and (1.1c) demonstrate how we can move directly from a balanced chemical equation, such as Reaction (1.1a), to a statement about the relative numbers of molecules (1.1b) or the relative numbers of moles (1.1c) involved in the reaction.

Exercise 1.1. An important chemical reaction in atmospheric, earth, and ocean sciences is that of dissolved carbon dioxide with liquid water to form carbonic acid, $H_2CO_3(l)$,

$$CO_2(g) + H_2O(l) \rightarrow H_2CO_3(l) \tag{1.2}$$

Calculate the mass of carbonic acid that forms for every kilogram of carbon dioxide that reacts with liquid water.

Solution. From the balanced chemical equation (1.2) we see that for every mole of carbon dioxide that reacts with water one mole of carbonic acid is formed. Since the molecular weight of CO_2 is 44.01, the number of moles of CO_2 in 1 kg is $1000/44.01 = 22.72$. Therefore, 22.72 moles of carbonic acid will form for every kilogram of CO_2 that reacts with water. The molecular weight of carbonic acid is 62.02, therefore, the number of grams of carbonic acid in 22.72 moles is $(22.72 \times 62.02) = 1409$. Therefore, for every kilogram of CO_2 that reacts with water 1.409 kg of carbonic acid are formed.

1.2 Equilibrium constants

A vapor is in equilibrium with its liquid when the rate of condensation is equal to the rate of evaporation. An analogous state of equilibrium exists in a chemical system when the rate at which the reactants combine to form products is equal to the rate at which the products decompose to form the reactants. For example, in the stratosphere and mesosphere, ozone (O_3) is formed by the reaction

$$O(g) + O_2(g) \rightarrow O_3(g) \tag{1.3}$$

However, some of the ozone molecules so formed break up again

$$O_3(g) \rightarrow O(g) + O_2(g) \tag{1.4}$$

Reaction (1.3) is called the *forward* reaction and Reaction (1.4) the *reverse* reaction. Reactions (1.3) and (1.4) can be combined as follows

$$O(g) + O_2(g) \rightleftarrows O_3(g)$$

At every temperature there exists partial pressures of the gases for which the forward and reverse reactions occur at the same rate; under these conditions, the system is said to be in *chemical equilibrium*.

A general chemical reaction can be represented by

$$aA + bB + \ldots \rightleftarrows gG + hH + \ldots \tag{1.5}$$

where A, B,... and G, H,... represent the chemical *reactants* and *products*, respectively, and $a,b,\ldots$ and $g,h,\ldots$ their coefficients in the balanced chemical equation. If Reaction (1.5) is in chemical equilibrium, and if the reactants and products are ideal gases or are present as solutes in an ideal solution,[4] then

$$\frac{[G]^g[H]^h\ldots}{[A]^a[B]^b\ldots} = K_c \qquad (1.6)$$

where [A], [B], ... and [G], [H], ... represent the equilibrium concentrations of the reactants and products, and K_c is called the *equilibrium constant* for the forward reaction (or, simply, the equilibrium constant for the reaction).[5] Equilibrium constants for some chemical reactions are given in Appendix IV. The value of K_c for a chemical reaction depends only on temperature (see Section 2.2), not on the concentrations of the chemical species or the volume or pressure of the system.

The concentrations in Eq. (1.6) may be expressed in *molarity* (M). For a gas the molarity is the number of *moles* of gas per liter of air; for a solution, it is the number of moles of solute per liter of solution. If any of the reactants or products are pure liquids or pure solids, their concentrations (i.e., densities) are essentially constant, compared to the large changes that are possible in the concentrations of the gases. Therefore, the concentrations of liquids and solids are incorporated into the value of K_c. The practical consequence of this is that the concentration of any pure liquid or pure solid may be equated to unity in Eq. (1.6).

Exercise 1.2. At 2000°C the value of K_c for the reaction

$$N_2(g) + O_2(g) \rightleftarrows 2NO(g) \qquad (1.7)$$

is 1.0×10^{-4}. If the equilibrium concentrations of $O_2(g)$ and $NO(g)$ are 50 M and 0.030 M, respectively, what is the equilibrium concentration of $N_2(g)$?

Solution. Application of Eq. (1.6) to Reaction (1.7) yields

$$K_c = \frac{[NO(g)]^2}{[N_2(g)][O_2(g)]}$$

Therefore,

$$1.0 \times 10^{-4} = \frac{(0.030)^2}{[N_2(g)](50)}$$

and,

$$[N_2(g)] = 0.18 \text{ M}$$

At 25°C the value of K_c for Reaction (1.7) is only 1×10^{-30}! This implies that the equilibrium concentration of NO(g) is very low at normal temperatures and that the equilibrium "lies to the left" of Reaction (1.7), favoring the reactants. Hence, in the troposphere, negligible quantities of NO(g) are produced by Reaction (1.7).

In the case of chemical reactions involving only gases, it is often more convenient to express the equilibrium constant for the reaction in terms of the partial pressures of the reactants and products instead of their molarities. However, before doing this we must review the ideal gas equation.

Laboratory experiments show that for a wide range of conditions the pressure (p), volume (V) and temperature (T) of all gases follow closely the same relationship, which is called the *ideal gas equation*. In SI units (see Appendix I), the ideal gas equation can be written in the following forms. For mass m (in kilograms) of a gas

$$pV = mRT \tag{1.8a}$$

where p is in pascals, V in cubic meters, T in K ($K = °C + 273.15 \simeq °C + 273$), and R is the gas constant for 1 kg of a gas. The value of R depends on the number of molecules in 1 kg of the gas, and therefore varies from one gas to another. Since $m/V = \rho$, where ρ is the density of the gas,

$$p = R\rho T \tag{1.8b}$$

For 1 kg of gas ($m = 1$), Eq. (1.8a) becomes

$$p\alpha = RT \tag{1.8c}$$

where α is the specific volume of the gas (i.e., the volume occupied by 1 kg of the gas). One mole of any gas contains the same number of molecules (N_A). Therefore, the gas constant for 1 mole is the same for all gases and is called the *universal gas constant* R^*(8.3143 J deg^{-1} mol^{-1}). Therefore,

$$pV = nR^*T \tag{1.8d}$$

where n is the number of moles of the gas, which is given by

$$n = \frac{1000m}{M}$$

where 1000 m is the number of grams of the gas and M the molecular weight of the gas. Also,

$$R^* = M\frac{R}{1000} \qquad (1.8e)$$

where R is divided by 1000 to obtain the gas constant for 1 g of gas. It can be seen from Eq. (1.8d) that at constant temperature and pressure the volume occupied by any gas is proportional to the number of moles (and therefore the number of molecules) in the gas. The gas constant for 1 molecule of any gas is also a universal constant, called the *Boltzmann constant* k. Since the gas constant for N_A molecules is R^*

$$k = \frac{R^*}{N_A} = \frac{8.3143}{6.022 \times 10^{23}} = 1.381 \times 10^{-23} \text{ J deg}^{-1} \text{ molecule}^{-1} \qquad (1.8f)^6$$

For a gas containing n_0 molecules per cubic meter, the gas equation can be written

$$p = n_0 kT \qquad (1.8g)$$

In chemistry, it is common because it is convenient, to depart from SI units in the gas equation and, instead, to express pressure in atmospheres and volume in liters (T is still in K). In this case, for n_A moles of gas A with pressure p_A and volume V_A we can write the ideal gas equation as

$$p_A V_A = n_A R_c^* T \qquad (1.8h)$$

where R_c^* is the universal gas constant in "chemical units" (indicated by the subscript c); the value of R_c^* is 0.0821 L atm deg^{-1} mol^{-1}. Since n_A/V_A is the number of moles of the gas per liter, that is, the molarity [A] of the gas

$$[A] = \frac{n_A}{V_A} = \frac{p_A}{R_c^* T} \qquad (1.8i)$$

Exercise 1.3. Carbon dioxide occupies about 354 parts per million by volume (ppmv) of air. How many CO_2 molecules are there in 1 m^3 of air at 1 atm and 0°C?

Solution. Let us calculate first the number of molecules in 1 m^3 of any gas at 1 atm and 0°C (which is called the *Loschmidt number*). This is given by n_0 in Eq. (1.8g) with $p = 1$ atm $= 1013 \times 10^2$ Pa, $T = 273$K and $k = 1.381 \times 10^{-23}$ J deg^{-1} molecule^{-1}. Therefore,

$$\text{Loschmidt number} = \frac{1013 \times 10^2}{(1.381 \times 10^{-23})273} = 2.69 \times 10^{25} \text{ molecule m}^{-3}$$

Since, at the same temperature and pressure, the volumes occupied by gases are proportional to the numbers of molecules in the gases, we can write

$$\frac{\text{Volume occupied by CO}_2 \text{ molecules in air}}{\text{Voume occupied by air}} = \frac{\text{Number of CO}_2 \text{ molecules in 1 m}^3 \text{ of air}}{\text{Total number of molecules in 1 m}^3 \text{ of air}}$$

Therefore,

$$354 \times 10^{-6} = \frac{\text{Number of CO}_2 \text{ molecules in 1 m}^3 \text{ of air}}{2.69 \times 10^{25}}$$

Hence, the number of CO_2 molecules in 1 m^3 of air is $(354 \times 10^{-6}) \times (2.69 \times 10^{25}) = 9.52 \times 10^{17}$.

We can now derive an expression for the equilibrium constant for a chemical reaction involving only gases in terms of the partial pressures of the gases. From Eqs. (1.6) and (1.8i)

$$K_c = \frac{[p_G/R_c^*T]^g[p_H/R_c^*T]^h \cdots}{[p_A/R_c^*T]^a[p_B/R_c^*T]^b \cdots} = \frac{p_G^g p_H^h \cdots}{p_A^a p_B^b \cdots}(R_c^*T)^{\Delta n}$$

or,

$$K_c = K_p(R_c^*T)^{\Delta n} \tag{1.9a}$$

where,

$$K_p = \frac{p_G^g p_H^h \cdots}{p_A^a p_B^b \cdots} \tag{1.9b}$$

and,

$$\Delta n = (a + b + \ldots) - (g + h + \ldots) \tag{1.9c}$$

K_p is generally used as the equilibrium constant in problems involving gaseous reactions. As in the case of K_c, terms for pure liquids and solids do not appear in the expression for K_p, and the coefficients for these terms are taken to be zero in the expression for Δn. Note that the units, as well as the numerical values, of K_c and K_p may differ. For example, for Reaction (1.3) the units of K_c are those of

$$\frac{[O_3(g)]}{[O(g)][O_2(g)]} \text{ or } \frac{M}{(M)(M)} = M^{-1}, \text{ the units of } K_p \text{ are}$$

$$\frac{p_{O_3}}{(p_O)(p_{O_2})} \text{ or } \frac{atm}{(atm)(atm)} = atm^{-1}.$$

Nevertheless, it is common practice in chemistry not to indicate the units of equilibrium constants, with the understanding that when K_c is used the concentrations are in molarity, and when K_p is used the partial pressures are in atmospheres.

Exercise 1.4. Ammonia, $NH_3(g)$, is produced commercially from the reaction of hydrogen, $H_2(g)$, and atmospheric nitrogen, $N_2(g)$, at high temperatures. If $H_2(g)$, $N_2(g)$, and $NH_3(g)$ attain equilibrium at 472°C when their concentrations are 0.12 M, 0.04 M, and 0.003 M, respectively, calculate the values of K_c and K_p for the reaction at 472°C.

Solution. The balanced chemical equation for the reaction is

$$3H_2(g) + N_2(g) \rightleftarrows 2NH_3(g)$$

Hence, from Eq. (1.6)

$$K_c = \frac{[NH_3(g)]^2}{[H_2(g)]^3[N_2(g)]} = \frac{(0.003)^2}{(0.12)^3(0.04)} = 0.1$$

From Eq. (1.9a)

$$K_c = K_p(R_c^*T)\Delta n$$

where $K_c = 0.1$, $R_c^* = 0.0821$ L atm deg^{-1}mol^{-1}, $T = 745K$ and, from Eq. (1.9c), $\Delta n = (3+1) - (2) = 2$. Therefore,

$$K_p = \frac{0.1}{(0.0821 \times 745)^2} = 3 \times 10^{-5}$$

1.3 Reaction quotient

If the general chemical reaction represented by Eq. (1.5) is not in equilibrium, we can still formulate a ratio of concentrations that has the same form as Eq. (1.6). This is called the *reaction quotient*, Q

$$Q = \frac{[G]^g[H]^h...}{[A]^a[B]^b...} \tag{1.10}$$

Clearly, if $Q = K_c$, the reaction is in chemical equilibrium. If $Q < K_c$, the reaction is not in equilibrium, and it will proceed in the forward direction until $Q = K_c$. If $Q > K_c$, the reaction will proceed in the reverse direction until $Q = K_c$.

Exercise 1.5. If 0.80 mole of $SO_2(g)$, 0.30 mole of $O_2(g)$, and 1.4 mole of $SO_3(g)$ simultaneously occupy a volume of 2 L at 1000K, will the mixture be in equilibrium? If not, in what direction will it proceed to establish equilibrium? Consider only the species $SO_2(g)$, $O_2(g)$, and $SO_3(g)$ in the reaction

$$2SO_2(g) + O_2(g) \rightleftarrows 2SO_3(g) \tag{1.11}$$

with $K_c = 2.8 \times 10^2$.

Solution. The reaction quotient for Reaction (1.11) is

$$Q = \frac{[SO_3(g)]^2}{[SO_2(g)]^2[O_2(g)]}$$

To evaluate the initial value of Q we must determine their initial molarities. These are: for $SO_2 = 0.80/2 = 0.40$ M, for $O_2 = 0.30/2 = 0.15$ M, and for $SO_3 = 1.4/2 = 0.70$ M. Hence,

$$Q = \frac{(0.70)^2}{(0.40)^2(0.15)} = 20$$

Since this value of Q is not equal to K_c (namely, 2.8×10^2), the initial mixture is not in equilibrium. Moreover, since $Q < K_c$, Reaction (1.11) will proceed in the forward direction.

Exercise 1.6. What are the equilibrium concentrations of $SO_2(g)$, $O_2(g)$, and $SO_3(g)$ in Exercise 1.5?

Solution. If y moles (or $\frac{y}{2}$ M) of $SO_3(g)$ are formed, it follows from Reaction (1.11) that y moles (or $\frac{y}{2}$ M) of $SO_2(g)$ and $\frac{y}{2}$ moles (or $\frac{y}{4}$ M) of $O_2(g)$ disappear. If this change establishes chemical equilibrium we have

Reaction	$2SO_2(g)$	+	$O_2(g)$	$\rightleftarrows$	$2SO_3(g)$
Initial concentrations	0.40 M		0.15 M		0.70 M
Change in concentrations	$-\frac{y}{2}$ M		$-\frac{y}{4}$ M		$\frac{y}{2}$ M
Equilibrium concentrations	$\left(0.40 - \frac{y}{2}\right)$ M		$\left(0.15 - \frac{y}{4}\right)$ M		$\left(0.70 + \frac{y}{2}\right)$ M

Hence the equilibrium constant for this reaction is

$$K_c = \frac{[SO_3(g)]^2}{[SO_2(g)]^2[O_2(g)]} = \frac{\left(0.70 + \frac{y}{2}\right)^2}{\left(0.40 - \frac{y}{2}\right)^2\left(0.15 - \frac{y}{4}\right)}$$

Since $K_c = 2.8 \times 10^2$ at 1000°C, we have

$$2.8 \times 10^2 = \frac{\left(0.70 + \frac{y}{2}\right)^2}{\left(0.40 - \frac{y}{2}\right)^2\left(0.15 - \frac{y}{4}\right)}$$

Rearranging and simplifying yields

$$-70y^3 + 153y^2 - 115y + 25 = 0$$

The reader may verify by substitution that an approximate solution to this cubic equation is $y = 0.37$. Hence, the equilibrium concentrations of $SO_2(g)$, $O_2(g)$, and $SO_3(g)$ are approximately 0.21 M, 0.057 M, and 0.89 M, respectively.

1.4 LeChatelier's principle

The way in which a system at equilibrium will respond to an imposed change can be predicted in a qualitative sense by *LeChatelier's principle,* which states that *if a system at equilibrium is subjected to a disturbance that changes any of the factors that determine its state of equilibrium, the system will react in such a way as to minimize (i.e., relieve) the effect of the disturbance.*

Before applying LeChatelier's principle to chemical systems, let us apply it to evaluate the effect of pressure on the melting point of ice. At a pressure of 1 atm the melting point of ice is 0°C; under these conditions liquid water and ice can coexist in equilibrium. According to LeChatelier's principle, if the pressure is increased the ice–water system will react in such a way as to tend to relieve the increase in pressure. Since the specific volume of water is less than that of ice, this is accomplished by the ice melting. Hence, when the pressure exceeds 1 atm, 0°C is no longer a sufficiently low temperature for ice and water to exist in equilibrium. In other words, the melting point of the system is lowered by applying pressure.

Now let us apply LeChatelier's principle to predict the effects on a chemical reaction of changing the concentrations of any of the species involved in the reaction. Consider, for example, the reaction between $H_2(g)$ and $N_2(g)$ to form $NH_3(g)$ at high temperature (see Exercise 1.4). If either $H_2(g)$ or $N_2(g)$ is added to an equilibrium system more $NH_3(g)$ will form, since by doing so the concentration of $H_2(g)$ or $N_2(g)$ will tend to be returned closer to its original concentration. Alternatively, if $NH_3(g)$ is added to the system, it will tend to decompose into $H_2(g)$ and $N_2(g)$.

Consider next the effect of changing the volume of a chemical system. For example, suppose there is a decrease in the volume in which the equilibrium Reaction (1.11) occurs but with temperature remaining constant. This will cause an increase in pressure; therefore, the system will react in such a way as to relieve this increase, which can be accomplished if the number of moles of gas in the system is reduced. Inspection of Reaction (1.11) shows that 3 moles of gases on the left produce 2 moles of gas on the right. Hence, when the volume of this system is reduced at constant temperature, more $SO_3(g)$ is produced.

Finally, let us apply LeChatelier's principle to determine the effect of temperature on an equilibrium chemical reaction. Raising the temperature of a system is equivalent to adding heat. Therefore, if the temperature of a system is raised at constant pressure, the chemical reaction will proceed in the direction that *absorbs* heat (i.e., in the direction of the so-called *endothermic* reaction). For example, the reverse reaction of Reaction (1.11) is endothermic, but the forward reaction is *exothermic* (i.e., heat is *released* by the forward reaction). Therefore, the equilibrium shifts in the forward direction if the temperature is lowered, and in the reverse direction if the temperature is raised at constant pressure.

Exercises

1.7. Answer, interpret, or explain the following in light of the principles presented in this chapter:

(a) Why do chemists prefer to use the mole as a unit of mass rather than, say, 1 kg?

(b) A gram–atomic weight can be defined in an analogous way to a gram–molecular weight. A gram–atomic weight of any element contains Avogadro's number of atoms of that element.

(c) In one gram–molecular weight of a compound there must be at least one gram–atomic weight of a given element, or some integral multiple of this weight. Does this suggest a method for determining atomic weights?

(d) The number of moles in any gas sample can be found by comparing its volume at STP with 22.4 L.

(e) If two elements form more than one compound, then the different masses of one that combine with the same mass of the other are in the ratio of small whole numbers (the *law of multiple proportions*).

(f) A chemical reaction generally starts off fast, slows with time, and finally ceases.

(g) The equilibrium constant for a chemical reaction is independent of the exact mechanism of the reaction, the rate at which the equilibrium is approached, or the direction from which the equilibrium is approached. (Compare with a mechanical system.)

(h) Only for reactions in which the number of molecules of gaseous reactants is different from the number of molecules of gaseous products does a volume change remove the system from equilibrium.

1.8. Write expressions for the equilibrium constants (K_c) for both the forward and reverse reactions of the following chemical equations:

(a) $\quad\quad\quad\quad N_2(g) + 3H_2(g) \rightleftarrows 2NH_3(g)$

(b) $\quad\quad\quad\quad N_2O(g) + \frac{1}{2}O_2(g) \rightleftarrows 2NO(g)$

(c) $\quad\quad\quad\quad 4NH_3(g) + 3O_2(g) \rightleftarrows 2N_2(g) + 6H_2O(g)$

(d) $\quad\quad\quad\quad NH_4Cl(s) \rightleftarrows NH_3(g) + HCl(g)$

1.9. If the equilibrium constant (K_c) for the reaction

$$A(g) + 2B(g) \rightleftarrows G(g) + 3H(g)$$

(where A, B, G, and H represent any chemical species) is 2.1×10^{-3}, what concentration of H(g) will be in equilibrium with 0.1 M of A(g), 0.25 M of B(g), and 0.02 M of G(g)?

1.10. The equilibrium constant (K_p) for the reaction

$$2SO_2(g) + O_2(g) \rightleftarrows 2SO_3(g)$$

at 727°C is 3.39. What is the partial pressure of $SO_3(g)$ that would be in equilibrium with 0.20 atm of $SO_2(g)$ and 0.80 atm of $O_2(g)$?

1.11. What is the value of K_c at 727°C for the reaction given in Exercise 1.10?

1.12. If 6.14 kg of $N_2O_4(g)$ equilibrate with 0.062 kg of $NO_2(g)$ at 25°C in a 6.00-L vessel, what is the value of K_c for the following reaction

$$N_2O_4(g) \rightleftarrows 2NO_2(g)$$

1.13. If 7.60 g of $N_2O_4(g)$ and 1.60 g of $NO_2(g)$ are mixed simultaneously within a volume of 3.00 L at 25°C, what will be the equilibrium masses of $N_2O_4(g)$ and $NO_2(g)$ established by the reaction given in Exercise 1.12 if the equilibrium constant K_c at 25°C is 4.56×10^{-3}?

1.14. Prove:

(a) The equilibrium constant for a reverse reaction is the reciprocal of the equilibrium constant for the corresponding forward reaction.

(b) If the coefficients in a balanced chemical equation are multiplied by a common factor (say m), the new equilibrium constant will be the old one raised to the power of m.

(c) The equilibrium constant for a net reaction is the product of the equilibrium constants for the individual reactions that add together to give the net reaction.

1.15. Equilibria involving $SO_2(g)$, $O_2(g)$ and $SO_3(g)$ can be described by

$$2SO_2(g) + O_2(g) \rightleftarrows 2SO_3(g) \qquad \text{(i)}$$

or,

$$2SO_3(g) \rightleftarrows 2SO_2(g) + O_2(g) \qquad \text{(ii)}$$

or,

$$SO_2(g) + \tfrac{1}{2}O_2(g) \rightleftarrows SO_3(g) \qquad \text{(iii)}$$

If the equilibrium constant for Reaction (i) is 2.8×10^2 at 1000K, what are the values of the equilibrium constants at 1000K for Reactions (ii) and (iii)?

1.16. At 823K, the equilibrium constants (K_c) for the reactions

$$COO(s) + H_2(g) \rightleftarrows CO(s) + H_2O(g)$$

and,

$$COO(s) + CO(g) \rightleftarrows CO(s) + CO_2(g)$$

are 67 and 490, respectively. What is the equilibrium constant at 823K for the following reaction?

$$CO_2(g) + H_2(g) \rightleftarrows CO(g) + H_2O(g)$$

1.17. If the equilibrium constant K_p for the reaction

$$NH_4Cl(s) \rightleftarrows NH_3(g) + HCl(g)$$

is 4×10^{-8} atm^2 at a temperature T, what will be the equilibrium vapor pressures of $NH_3(g)$ and $HCl(g)$ produced by the decomposition of $NH_4Cl(s)$ at temperature T?

1.18. At the high temperatures at which exhaust gases are emitted from automobiles, the reaction

$$2CO_2(g) \rightleftarrows 2CO(g) + O_2(g)$$

has an equilibrium constant K_p of about 1×10^{-13} atm. If the percentages by volume of the exhaust gases at 1 atm are 0.30 of $CO(g)$, 13 of $CO_2(g)$, and 4.0 of $O_2(g)$, what is the reaction quotient for the above reaction? As the reaction moves to equilibrium, what gases will be produced?

1.19. By applying the concepts of chemical equilibrium constants to the physical equilibrium between liquid water and water vapor

$$H_2O(l) \rightleftarrows H_2O(g)$$

show that the equilibrium vapor pressure of water is given by

$$p_{H_2O} = [H_2O(g)] \, R_c^* T$$

where $[H_2O(g)]$ is the concentration in molarity of the $H_2O(g)$. (Note that this relation is a form of the ideal gas equation.)

1.20. If Reaction (c) given in Exercise 1.8 is at equilibrium, what will be the effect on the amount of water vapor of decreasing the volume of the system at constant temperature?

1.21. If in the reaction

$$N_2O_4(g) \rightleftarrows 2NO_2(g)$$

a fraction f of the original concentration of $N_2O_4(g)$ has dissociated when equilibrium is reached, derive an expression for the equilibrium constant K_p for the reaction in terms of f and the total pressure p of the system at equilibrium.

1.22. Solid ammonium mercaptan, $NH_4HS(s)$, dissociates rapidly at room temperature to form ammonia and hydrogen sulfide

$$NH_4HS(s) \rightleftharpoons NH_3(g) + H_2S(g)$$

where $K_p = 1.08 \times 10^{-1}$ at 25°C. If some $NH_4HS(s)$ is placed in a closed 2.00-L flask that already contains 0.300 g of $NH_3(g)$, what will be the total pressure in the flask after chemical equilibrium is established?

1.23. In automobile engines, why is some $NO(g)$ produced by the following endothermic reaction?

$$N_2(g) + O_2 \rightarrow 2NO(g)$$

1.24. With reference to Exercise 1.23, why does not most of the $NO(g)$ that is produced quickly revert to $N_2(g)$ and $O_2(g)$ when the emissions attain normal atmospheric temperatures?

Notes

1 The phase of a substance is indicated in the parenthesis following the chemical symbol, where "g" stands for gas, "l" for liquid, "s" for solid, and "aq" for aqueous (i.e., water solution) phase.

2 If this is not intuitively obvious, it can be proved as follows. Let the masses of the molecules of two compounds 1 and 2 be m_1 and m_2 and their molecular weights M_1 and M_2, respectively. Then, by definition, 1 mole of compound 1 is M_1 g of compound 1, and 1 mole of compound 2 is M_2 g of compound 2. Let the numbers of molecules in M_1 g of compound 1 and M_2 g of compound 2 be n_1 and n_2, respectively. Then, $M_1 = n_1 m_1$ and $M_2 = n_2 m_2$. Therefore, $M_1/M_2 = n_1 m_1/n_2 m_2$. However, the ratio of the molecular weights of any compounds is equal to the ratio of the masses of their molecules, that is, $M_1/M_2 = m_1/m_2$. It follows that $n_1 = n_2$, that is, the number of molecules in 1 mole of compound 1 is equal to the number of molecules in 1 mole of compound 2.

3 Prior to 1982, the standard pressure was one atmosphere (1 atm), and this is still in common use. The difference in the two definitions is not great since 1 atm = 1.013 bar. The unit of pressure in the International System of Units (i.e., the *SI system*) is the *pascal* (Pa). The basic and derived units for the SI system are given in Appendix I; for the most part, we will adhere to the SI system in this book.

4 A *solution* is a homogeneous mixture. For example, air is a gaseous solution of several gases, seawater is a liquid solution of sodium chloride and other materials. The component of a solution that is present in the greatest amount, and therefore determines the state of matter (solid, liquid or gas) of the solution, is called the *solvent;* the other components are called *solutes*. A solution in which water is the solvent (e.g., seawater) is called an *aqueous solution*. An *ideal solution* is one for which both solvent and solutes obey Raoult's law (see Section 4.4) at all concentrations.

5 For reactions between gases at high pressures, or for reactions in nonideal solutions, K_c defined by Eq. (1.6) is not strictly constant. In these cases, a thermodynamic

equilibrium constant is defined in terms of the *activities* of the reactants and products. We will not consider this complication here since, for gases at pressure < 1 atm, K_c is essentially constant.

6 The numerical value of k given here is expressed to an accuracy of four significant figures. This is because this is the number of significant figures in the *least* accurate numerical value that we have used to derive the value of k (namely, 6.022×10^{23}). This same principle will be followed throughout this text, and in solving numerical problems.

2

Chemical thermodynamics

Heat can be released or absorbed during a chemical reaction. This provides a powerful method for studying chemical equilibrium by means of chemical thermodynamics. Thermodynamics is based on a few fundamental postulates, called the *first, second,* and *third laws of thermodynamics.* We will discuss these laws first, and then return to the subject of chemical equilibrium.

2.1 The first law of thermodynamics; enthalpy

In addition to the macroscopic kinetic and potential energy that a body or system as a whole may possess, it also contains *internal energy* due to the kinetic and potential energy of its molecules or atoms. Increases in internal kinetic energy in the form of molecular motions are manifested as increases in the temperature of the system, while changes in the potential energy of the molecules are caused by changes in their relative configurations.

Let us suppose that a system of unit mass takes in a certain quantity of heat energy q (measured in joules). As a result, the system may do a certain amount of external work w (also measured in joules). The excess energy supplied to the system, over and above the external work done by the system, is $q - w$. Therefore, if there is no change in the macroscopic kinetic and potential energy of the system, it follows from the principle of conservation of energy that the internal energy of the system must increase by $q - w$. That is,

$$q - w = u_2 - u_1 \tag{2.1}$$

where u_1 and u_2 are the internal energies of a unit mass of the system before and after the change. In differential form Eq. (2.1) becomes

$$dq - dw = du \qquad (2.2)$$

where dq is the differential increment of heat added to a unit mass of the system, dw the differential increment of work done by a unit mass of the system, and du the differential increment in internal energy of a unit mass of the system. Equations (2.1) and (2.2) are statements of the *first law of thermodynamics*. In fact, Eq. (2.2) provides a definition of du. It should be noted that the change in internal energy du is a function only of the initial and final states of the system and is therefore independent of the manner by which the system is transferred between these two states. Thermodynamic variables that possess this property are called *functions of state*. For example, pressure volume and temperature are functions of state.

To visualize the work term dw in Eq. (2.2) in a simple case, consider a substance (often called the *working substance*) contained in a cylinder of fixed cross-sectional area which is fitted with a movable, frictionless piston (Fig. 2.1). The volume of the substance is then proportional to the distance from the base of the cylinder to the face of the piston, and can be represented on the horizontal line of the graph shown in Figure 2.1. The pressure of the substance in the cylinder can be represented on the vertical line of this graph. Therefore, every state of the substance corresponding to a given position of the cylinder is represented by a point of the graph. When the substance is in equilibrium at a state represented by the point P on this graph, its pressure is p and its volume V. If the piston moves outward through an incremental distance dx, while the pressure remains essentially constant at p, the work dW done by the substance in expanding is equal to the force exerted on the piston (this force is equal to pA where A is the cross-sectional area of the piston) multiplied by the distance dx through which the piston moves. That is,

$$dW = pA \, dx = p \, dV \qquad (2.3)$$

In other words, the work done by the substance when its volume increases by a small amount is equal to the pressure of the substance multiplied by its increase in volume. It should be noted that $dW = p \, dV$ is equal to the shaded area in the graph shown in Figure 2.1; that is, it is equal to the area under the curve PQ. When the substance passes from state A with volume V_1 to state B with volume V_2 (Fig. 2.1), during which its pressure p changes, the work W done by the substance is equal to the area under the curve AB. That is,

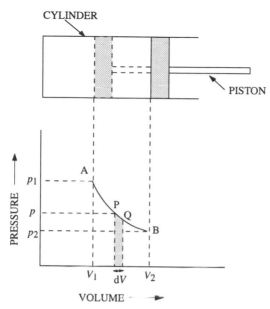

Figure 2.1. Representation of the state of a substance in a cylinder on a
p-V diagram.

$$W = \int_{V_1}^{V_2} p \, dV \tag{2.4}$$

Equations (2.3) and (2.4) are quite general and represent the work
done by any substance or system when its volume changes.

If we are dealing with a unit mass of a substance, the volume V is
replaced by the specific volume α (i.e., volume per unit mass) and the
work dw that is done by a unit mass of the substance when its specific
volume increases by $d\alpha$ is

$$dw = p \, d\alpha \tag{2.5}$$

Combining Eqs. (2.2) and (2.5) yields

$$dq = du + p \, d\alpha \tag{2.6}$$

which is an alternative statement of the first law of thermodynamics.
It should be noted that the first law of thermodynamics can be applied
to any system: chemical or physical, solid, liquid, or gas, or to any
combination of such systems.

If heat is added to a system at constant pressure, so that the specific
volume of the system increases from α_1 to α_2, the work done by a unit

mass of the system is $p(\alpha_2 - \alpha_1)$. Therefore, from Eq. (2.6), the heat dq added to a unit mass of the system at constant pressure is given by

$$dq = (u_2 - u_1) + p(\alpha_2 - \alpha_1) = (u_2 + p\alpha_2) - (u_1 + p\alpha_1)$$

where u_1 and u_2 are, respectively, the initial and final internal energies for unit mass. Therefore, at constant pressure,

$$dq = h_2 - h_1 = dh \tag{2.7}$$

where h is the *enthalpy* (sometimes called the *heat at constant pressure*) of a unit mass of the system, which is defined by

$$h = u + p\alpha \tag{2.8}$$

If the quantity of heat (in joules) required to raise a unit mass of the system by 1°C at constant pressure (called the *specific heat at constant pressure*) is c_p, then

$$dq = c_p \, dT \tag{2.9}$$

From Eqs. (2.7) and (2.9)

$$dh = c_p \, dT \tag{2.10}$$

or, on integrating,

$$h = c_p \, T \tag{2.11}$$

where the value of h is taken to be zero at the absolute zero of temperature ($T = 0K$).

   Exercise 2.1. A parcel of dry air at 1 atm pressure receives 10^7 J of heat by radiation from the sun, and its volume increases by 22 m³. If the center of the mass of the parcel does not move, what is the change in the internal energy of the parcel? If the molecules in the air exert no forces on each other, what is the increase in the temperature of the air parcel if it has a mass of 8000 kg? (Specific heat at constant pressure of dry air $= 1004$ J deg^{-1} kg^{-1}.)

   Solution. Since the center of mass of the parcel does not move, there is no change in either the macroscopic potential energy or the kinetic energy of the parcel. Also, the pressure of the system remains constant at 1 atm $= 1.013$ bar $= 1.013 \times 10^5$ Pa. Therefore, the first law of thermodynamics in the form of Eq. (2.6) applies, which, for a system as a whole (rather than a unit mass), can be written

$$dQ = dU + pdV \tag{2.12}$$

where dQ is the heat added to the system, dU the increase in internal energy of the system, and pdV the work done by the system when its volume increases by dV. For the air parcel $dQ = 10^7$ J and $pdV = (1.013 \times 10^5)(22) = 0.22 \times 10^7$ J. Therefore,

$$dU = dQ - pdV = 10^7 - 0.22 \times 10^7 = 7.8 \times 10^6 \text{ J}$$

If the molecules in the air exert no forces on each other, the internal energy of the air must consist entirely of the kinetic energy of the molecules, that is, on the temperature of the air.[1] From the definition of c_p, the energy required to raise a mass m of a system by $\Delta T °C$ at constant pressure is $mc_p \Delta T$. Hence, $dU = mc_p \Delta T$ or $\Delta T = dU/mc_p$. Substituting $dU = 7.8 \times 10^6$ J, $m = 8 \times 10^3$ kg and $c_p = 1004$ J deg^{-1} kg^{-1} into this expression yields the temperature rise of the air parcel, namely, $\Delta T = 0.97 °C$.

2.2 Enthalpies of reaction and formation

If the temperature is kept constant, changes in the concentrations of chemical species or changes in the volume or pressure of a system do not change the equilibrium constants K_c or K_p. However, changes in temperature do change the equilibrium constants. These changes can be represented by

$$K = A \exp \left(\frac{-\Delta \overline{H}^0_{rx}}{R^* T} \right) \tag{2.13}$$

where $\Delta \overline{H}^0_{rx}$ is called the *molar standard enthalpy (or heat) of reaction*, R^* is the universal gas constant, and T is the temperature (in K). A and $\Delta \overline{H}^0_{rx}$ are constants (over a reasonable temperature range) for any given chemical reaction. The line above the H indicates that the molar amounts of the reactants and products given by the numerical coefficients in the balanced chemical equation are involved. The superscript zero to H indicates that the reactants and products must be in their standard states, which are generally defined to be the chemical forms most stable at 1 atm and 25°C.[2] For the forward reaction of the general chemical reaction (1.5), $\Delta \overline{H}^0_{rx}$ is given by

$$\Delta \overline{H}^0_{rx} = [\text{g } \Delta \overline{H}^0_f(G) + \text{h } \Delta \overline{H}^0_f(H) + ...] - [\text{a } \Delta \overline{H}^0_f(A) + \text{b } \Delta \overline{H}^0_f(B) + ...] \tag{2.14}$$

where $\Delta \overline{H}^0_f(X)$ is the difference in enthalpy between one mole of compound X in its standard state and its elements in their standard states. $\Delta \overline{H}_f^0(X)$ is called the *molar standard enthalpy (or heat) of formation* (or simply the *molar heat of formation*) of compound X. Some values of $\Delta \overline{H}_f^0$ are given in Appendix V. For an endothermic reaction $\Delta \overline{H}_{rx}^0$ is positive, and for an exothermic reaction it is negative. Similarly, positive values of $\Delta \overline{H}_f^0$ indicate that heat is absorbed when a compound is formed from its constituent elements in their standard states, and negative values of $\Delta \overline{H}_f^0$ indicate that heat is released. By convention, a value of zero is assigned to the enthalpies of formation of the elements in their standard states. The following example should clarify these points.

Exercise 2.2. Calculate $\Delta \overline{H}_{rx}^0$ for the combustion of ethane, $C_2H_6(g)$:

$$2C_2H_6(g) + 7O_2(g) \rightarrow 4CO_2(g) + 6H_2O(l) \tag{2.15}$$

Solution. From Appendix V we see that the $\Delta \overline{H}_f^0$ values for $C_2H_6(g)$, $CO_2(g)$, and $H_2O(l)$ are -84.7, -393.5, and -285.6 kJ mol^{-1}. Applying Eq. (2.14) to Reaction (2.15) yields

$$\Delta \overline{H}_{rx}^0 = \{4\Delta \overline{H}_f^0(CO_2(g)) + 6\Delta \overline{H}_f^0(H_2O(l))\} - 2\Delta \overline{H}_f^0(C_2H_6(g))$$
$$= \{4(-393.51) + 6(-285.85) - 2(-84.68)\} \text{ kJ}$$
$$= -3120 \text{ kJ}$$

The minus sign indicates that the reaction is exothermic (i.e., it releases heat). The enthalpy of formation of $O_2(g)$ does not appear in the expression for $\Delta \overline{H}_{rx}^0$ because $O_2(g)$ is, by definition, the stable form of elemental oxygen; therefore, its enthalpy of formation is zero. Note also that the value of $\Delta \overline{H}_{rx}^0$ that we have calculated applies specifically to Reaction (2.15) in which 2 moles of $C_2H_6(g)$ react with 7 moles of $O_2(g)$ to form 4 moles of $CO_2(g)$ and 6 moles of $H_2O(l)$. An alternate way of expressing the result would be to state that the enthalpy of the reaction is $-3120/2$ or -1560 kJ per mole of ethane burnt.

2.3 Entropy and the second law of thermodynamics

The incremental change in the entropy (ds) of a unit mass of a system is defined as

$$ds = \frac{dq_{rev}}{T} \tag{2.16}$$

or, for a finite change from state 1 to state 2,

$$\Delta s = \int_1^2 \frac{dq_{rev}}{T} \qquad (2.17)$$

where, dq_{rev} is the quantity of heat added *reversibly* to a system at temperature T. A *reversible* (or *equilibrium*) *transformation* is one in which a system moves by infinitesimal amounts and infinitesimally slowly between equilibrium states, so that the direction of the process can be reversed at any time just by making an infinitesimal change in the surroundings. Entropy is a function of state.

The second law of thermodynamics for a reversible transformation states (in part) that *for a reversible transformation there is no change in the entropy of the universe* (where "universe" refers to a system and its surroundings). In other words, if a system receives heat reversibly, the increase in its entropy is exactly equal in magnitude to the decrease in the entropy of its surroundings.

The concept of reversibility is an abstraction. All natural transformations are, in fact, irreversible. In an *irreversible* (or *spontaneous*) *transformation* a system undergoes finite changes at finite rates, and these changes cannot be reversed simply by changing the surroundings of the system by infinitesimal amounts.

Exercise 2.3. Prove that for the same change of state of a system, one carried out reversibly and the other irreversibly

$$w_{irrev} < w_{rev} \quad \text{and} \quad q_{irrev} < q_{rev}$$

where w_{irrev} and w_{rev} are the works of expansion done by a unit mass of a system during irreversible and reversible transformations, respectively, and q_{irrev} and q_{rev} are the corresponding quantities of net heat taken in by the system.

Solution. In a reversible transformation, state functions of a system (such as pressure) never differ from those of the surroundings by more than an infinitesimal amount. Therefore,

$$p_{system} = p_{surroundings} + dp$$

Hence, if a system expands reversibly, and in so doing passes from state 1 to state 2, the work of expansion done by a unit mass of the system is

$$w_{rev} = \int_1^2 p_{surroundings} d\alpha = \int_1^2 (p_{system} - dp) d\alpha \approx \int_1^2 p_{system} d\alpha$$

where α is the specific volume of the system. On the other hand, for an irreversible expansion between the same two states, $p_{\text{surroundings}} < p_{\text{system}}$. Therefore,

$$w_{\text{irrev}} = \int_1^2 p_{\text{surroundings}} d\alpha < \int_1^2 p_{\text{system}} d\alpha$$

Hence,

$$w_{\text{irrev}} < w_{\text{rev}} \qquad (2.18)$$

Applying the first law of thermodynamics in the form of Eq. (2.1) to the reversible transformation

$$q_{\text{rev}} - w_{\text{rev}} = u_2 - u_1$$

Similarly, for the irreversible transformation

$$q_{\text{irrev}} - w_{\text{irrev}} = u_2 - u_1$$

Therefore,

$$q_{\text{rev}} - w_{\text{rev}} = q_{\text{irrev}} - w_{\text{irrev}}$$

or,

$$q_{\text{rev}} - q_{\text{irrev}} = w_{\text{rev}} - w_{\text{irrev}}$$

Since, from Eq. (2.18), $w_{\text{rev}} - w_{\text{irrev}} > 0$, it follows that

$$q_{\text{rev}} - q_{\text{irrev}} > 0$$

or,

$$q_{\text{irrev}} < q_{\text{rev}} \qquad (2.19)$$

If a system receives heat dq_{irrev} at temperature T during an irreversible transformation, the change in the entropy of the system is *not* equal to dq_{irrev}/T. Also, for an irreversible transformation there is no simple relationship between the change in the entropy of the system and the change in the entropy of its surroundings. However, the remaining part of the second law of thermodynamics states that *the entropy of the universe increases as a result of irreversible transformations*.

The two parts of the second law of thermodynamics stated above can be summarized as follows:

$ds_{universe} = ds_{system} + ds_{surroundings}$ (2.20a)

$ds_{universe} = 0$ for reversible (equilibrium) transformations (2.20b)

$ds_{universe} > 0$ for irreversible (spontaneous) transformations (2.20c)

The second law of thermodynamics cannot be proved. It is believed to be valid because it leads to deductions that are in accord with observations and experience. The following exercise provides an example of such a deduction.

Exercise 2.4. Assuming the truth of the second law of thermodynamics, prove that an isolated ideal gas can spontaneously expand but not spontaneously contract.

Solution. We will consider a unit mass of the gas. If the gas is isolated, it has no contact with its surroundings; therefore, $ds_{surroundings} = 0$, and

$$ds_{universe} = ds_{system} + ds_{surroundings} = ds_{gas} \qquad (2.21)$$

Also, if the gas is isolated, $dq = dw = 0$; therefore, from Eqs. (2.5) and (2.6), $du = 0$. If $du = 0$, it follows from Joule's law for an ideal gas (see Note 1 in this chapter) that $dT = 0$. Hence, the gas must pass from its initial state (1) to its final state (2) isothermally.

To obtain an expression for ds_{gas}, we can follow any reversible and isothermal path from state 1 to state 2, and evaluate the integral

$$ds_{gas} = \int_1^2 \frac{dq_{rev}}{T}$$

For an ideal gas (see Exercise 2.28)

$$\frac{dq_{rev}}{T} = c_p \frac{dT}{T} - R \frac{dp}{p}$$

where R is the gas constant for a unit mass of the gas and c_p the specific heat at constant pressure of the gas. Therefore,

$$ds_{gas} = c_p \int_1^2 \frac{dT}{T} - R \int_1^2 \frac{dp}{p}$$

or,

$$ds_{gas} = c_p \ln \frac{T_2}{T_1} - R \ln \frac{p_2}{p_1}$$

Since $T_1 = T_2$, the ideal gas equation reduces to Boyle's law, which can be written as $p_1\alpha_1 = p_2\alpha_2$, where the α's are specific volumes. Therefore, the last expression becomes

$$ds_{gas} = -R \ln\frac{p_2}{p_1} = -R \ln\frac{\alpha_1}{\alpha_2} = R \ln\frac{\alpha_2}{\alpha_1} \qquad (2.22)$$

From Eqs. (2.21) and (2.22),

$$ds_{universe} = R \ln\frac{\alpha_2}{\alpha_1} \qquad (2.23)$$

Hence, if the second law of thermodynamics is valid, it follows from Eqs. (2.20c) and (2.23) that

$$R \ln\frac{\alpha_2}{\alpha_1} > 0$$

or,

$$\alpha_2 > \alpha_1$$

That is, the gas spontaneously expands. If, on the other hand, the gas spontaneously contracted, $\alpha_2 < \alpha_1$ and $ds_{universe} < 0$, which would violate the second law.

2.4 The third law of thermodynamics; absolute entropies

Although thermodynamics makes no assumptions about the structure of matter, it is sometimes instructive to interpret thermodynamic results in terms of microscopic properties. For example, entropy may be considered as a measure of the degree of disorder of the elements of a system: the more disorder the greater the entropy. From a molecular viewpoint, the disorder is associated with the molecules of a system.

As in the case of energy or enthalpy, we are usually interested in differences in entropy rather than absolute values. However, in the case of entropy, it is possible to assign absolute values. This is a consequence of the third law of thermodynamics, which states that *the entropy of perfect crystals of all pure elements and compounds is zero at the absolute zero of temperature (0 K)*. Consequently, the absolute entropy of a substance at any temperature T is given by the change in the entropy of the substance in moving from 0 K to T. The absolute entropies of many substances (generally at 25°C and 1 atm – indicated by $\overline{S}^0$ for the molar absolute entropy under standard conditions) are listed in chemical tables (a selection of $\overline{S}^0$ values is given in Appendix V).

The entropy changes associated with the forward reaction of the general chemical reaction (1.5) is given by

$$\Delta\overline{S}^0 = [g\overline{S}^0(G) + h\overline{S}^0(H) + ...] - [a\overline{S}^0(A) + b\overline{S}^0(B) + ...] \tag{2.24}$$

The following exercise illustrates the procedure.

Exercise 2.5. Calculate the change in entropy $\Delta\overline{S}^0$ at 298K associated with the reaction

$$2SO_2(g) + O_2(g) \rightarrow 2SO_3(g)$$

Solution. From Appendix V we see that the absolute molar entropies $\overline{S}^0$ of $SO_2(g)$, $O_2(g)$, and $SO_3(g)$ at 298K are 248.1, 205.0, and 256.1 J mol^{-1} deg.$^{-1}$ Therefore, from Eq. (2.24)

$$\Delta\overline{S}^0 = \{2\overline{S}^0[SO_3(g)] - 2\overline{S}^0[SO_2(g)] - \overline{S}^0[O_2(g)]\}$$
$$= \{2(256.1) - 2(248.1) - (205.0)\} \text{ J deg}^{-1}$$
$$\Delta\overline{S}^0 = -189.0 \text{ J deg}^{-1}$$

Note that the reaction decreases the number of molecules and the entropy decreases. In general, if a chemical reaction decreases the number of gaseous molecules, it decreases the entropy of a system; conversely, if a chemical reaction increases the number of gaseous molecules, it increases the entropy of the system.

2.5 Criteria for equilibrium and spontaneous transformation

If a system is in equilibrium with its surroundings, every possible infinitesimal transformation is reversible. Hence, a necessary condition for equilibrium is that Eq. (2.20b) holds for all infinitesimal transformations; that is, the sum of the entropy of the system and its surroundings is constant. This is the most general criterion for a system to be in equilibrium. Similarly, the most general criterion for a spontaneous transformation is given by Eq. (2.20c); that is, the transformation must result in an increase in the sum of the entropy of the system and its surroundings. However, these criteria are difficult to apply in practice because they involve the system and its surroundings, rather than the system alone.

To develop criteria for equilibrium and spontaneous transformations that involve only the system, we introduce a new function of state, called the *Gibbs free energy* (G, or g for unit mass) which is defined by

$$g = h - Ts \qquad (2.25)$$

or, using Eq. (2.8),

$$g = u + p\alpha - Ts \qquad (2.26)$$

Differentiating Eqs. (2.25) and (2.26)

$$dg = dh - Tds - sdT = du + pd\alpha + \alpha dp - Tds - sdT \qquad (2.27)$$

If we now confine ourselves to transformations at constant temperature and pressure (which are common conditions for chemical reactions)

$$dg = dh - Tds = du + pd\alpha - Tds$$

or, using Eq. (2.6),

$$dg = dh - Tds = dq - Tds \qquad (2.28)$$

Expressed in terms of finite changes in molar quantities for a chemical reaction, Eq. (2.28) becomes

$$\Delta \overline{G} = \Delta \overline{H}_{rx} - T \Delta \overline{S} \qquad (2.29)$$

which is called the *Gibbs–Helmholtz equation*.

For a system at constant pressure and temperature, we have from Eqs. (2.16) and (2.28)

$$dg = dq - dq_{rev} \qquad (2.30)$$

Now, if the transformation occurs under equilibrium conditions, and is therefore reversible, $dq = dq_{rev}$ and

$$dg = 0 \qquad (2.31)$$

If, on the other hand, the transformation is spontaneous, and therefore irreversible, it follows from Eq. (2.19) that $dq < dq_{rev}$. Hence, Eq. (2.30) becomes

$$dg < 0 \qquad (2.32)$$

Equation (2.31) shows that at equilibrium g must have a stationary value. We can establish whether this stationary value is a maximum or a minimum by utilizing Eq. (2.32). Since a spontaneous change must move a system toward equilibrium and $dg < 0$, the variation of g with the state of the system must be as shown in Figure 2.2, for, in this case, whether the system approaches the equilibrium point R from P

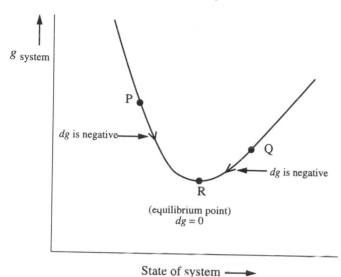

Figure 2.2. Sketch illustrating that, for a system at constant temperature and pressure, $dg = 0$ at the equilibrium point R and $dg < 0$ for spontaneous changes that move the system toward equilibrium (e.g., P to R or Q to R).

or from Q, dg is negative. That is, for a system at constant temperature and pressure at equilibrium, the Gibbs free energy has a minimum value.

It follows from Eqs. (2.28) and (2.32) that a spontaneous transformation is favored by a negative value of dq (i.e., when heat is given by a system to its surroundings) and by a positive value of ds (i.e., an increase in the entropy of a system).

These results can be summarized as follows. *The transformation of a system from one state to another, at constant temperature and pressure, is spontaneous if the Gibbs free energy (of the system alone) decreases. If the Gibbs free energy (of the system alone) is unchanged by the transformation, the two states are in equilibrium.* In other words, the criterion for the thermodynamic equilibrium of a system at constant temperature and pressure is that the Gibbs free energy of the system be at a minimum value. The Gibbs free energy is sometimes called the *thermodynamic potential at constant pressure* (constant temperature understood) in order to indicate its analogy with the potential energy of a mechanical system, which also has a minimum value under equilibrium conditions.

Note that for equilibrium (i.e., reversible) transformations at constant temperature and pressure (e.g., phase transitions at atmospheric pressure), we have from Eqs. (2.28) and (2.31)

$$ds = \frac{dh}{T} \qquad (2.33)$$

2.6 Standard free energy changes

The Gibbs free energy of a substance depends upon its state. By convention, the following are considered *standard states:* (1) for a solid, the pure solid at 1 atm and 25°C; (2) for a liquid, the pure liquid at 1 atm pressure and 25°C; (3) for a gas, an ideal gas at 1 atm partial pressure and 25°C; and (4) for a solution, an ideal solution with a concentration of 1 mole of solute per liter of solution (i.e., 1 M) at 25°C.

The change in Gibbs free energy of a system, when reactants in their standard states are converted to products in their standard states, is called the *molar standard free energy change* ($\Delta \overline{G}^0$) for the reaction. The superscript zero to the G indicates the standard state and the overbar indicates that the molar amounts of the reactants and products given by the numerical coefficients in the balanced chemical equation for the reaction are involved. For the forward reaction of the general chemical reaction (1.5)

$$\Delta \overline{G}^0 = [g\ \Delta \overline{G}_f^0(G) + h\ \Delta \overline{G}_f^0(H) + \ldots] - [a\ \Delta \overline{G}_f^0(A) + b\ \Delta \overline{G}_f^0(B) + \ldots] \qquad (2.34)$$

where $\Delta \overline{G}_f^0(X)$, which is called the *molar standard Gibbs free energy of formation* (or simply, the *standard free energy of formation*) of compound X, is the change in the Gibbs free energy when 1 mole of X is formed from its elements. By convention, the standard free energies of formation of the elements in their most stable forms at 1 atm are taken to be zero. The temperature chosen for tabulating values of $\Delta \overline{G}_f^0(X)$ is usually 25°C. A selection of standard free energies of formation is given in Appendix V.

It follows from the above definitions and Eq. (2.32) that if $\Delta \overline{G}_f^0(X)$ for a reaction is negative, the reactants in their standard states will be converted spontaneously into the products in their standard states. If $\Delta \overline{G}^0$ is positive, the conversion will not be spontaneous, but the corresponding reverse reaction will be. However, even when $\Delta \overline{G}^0$ is positive, some products can form but in concentrations below that of their standard states.

Exercise 2.6. Calculate $\Delta \overline{G}^0$ at 25°C and 1 atm for the reaction

$$H_2O_2(g) \rightarrow H_2O(g) + \tfrac{1}{2}O_2(g)$$

Is $H_2O_2(g)$ stable at 25°C and 1 atm?

Solution. From Appendix V we see that at 25°C and 1 atm $\Delta \overline{G}_f^0[H_2O_2(g)] = -188.2$ kJ mol^{-1} and $\Delta \overline{G}_f^0[H_2O(g)] = -228.6$ kJ mol^{-1} and, since $O_2(g)$ is the stable form of oxygen at 25°C and 1 atm, $\Delta \overline{G}_f^0[O_2(g)] = 0$. Hence, applying Eq. (2.34) to the above reaction

$$\Delta \overline{G}^0 = [(-228.6)] - [(-188.2)] \text{ kJ}$$

or,

$$\Delta \overline{G}^0 = -40.4 \text{ kJ}$$

Since $\Delta \overline{G}^0$ is negative, the reaction will proceed spontaneously in the forward direction and will be unstable. However, thermodynamic calculations give no information on the *speed* of chemical reactions. We will consider this subject in Chapter 3.

2.7 Free energy change and the equilibrium constant

It is only under standard conditions (1 atm and 25°C) that $\Delta \overline{G}^0$ is sufficient to determine whether a reaction is spontaneous or not. We will now derive a relation for the molar free energy change $\Delta \overline{G}$ under any conditions in terms of $\Delta \overline{G}^0$ and the equilibrium constant for the chemical reaction.

For a reversible (equilibrium) transformation, Eqs. (2.6), (2.16), and (2.27) can be combined to give

$$dg = \alpha dp - s \, dT \tag{2.35}$$

or at constant temperature,

$$dg = \alpha dp \tag{2.36}$$

Applied to n moles of a substance, Eq. (2.36) becomes

$$dG = nd\overline{G} = nVdp \tag{2.37}$$

where V is the volume of 1 mole of the substance. If we now consider 1 mole of an ideal gas, from the gas equation [see Eq. (1.8d)]

$$pV = R^*T \tag{2.38}$$

From Eqs. (2.37) and (2.38)

$$dG = \frac{R^*T}{p} dp \tag{2.39}$$

Integrating Eq. (2.39) between the pressure limits p° ($= 1$ atm) and p (in atm)[3]

$$\int_{\overline{G}_T^0}^{\overline{G}} d\overline{G} = \int_{p_0}^{p} \frac{R^*T}{p} dp$$

or,

$$\overline{G} - \overline{G}_T^0 = R^*T \ln \frac{p}{p^\circ} = R^*T \ln p \tag{2.40}$$

where $\overline{G}$ is the Gibbs free energy per mole at pressure p and temperature T and $\overline{G}_T^0$ the Gibbs free energy at 1 atm and temperature T.

For the general chemical reaction (1.5)

$$\Delta \overline{G} = [g\overline{G}(G) + h\overline{G}(H) + \ldots] - [a\overline{G}(A) + b\overline{G}(B) + \ldots]$$

Using Eq. (2.40) in this last expression

$$\Delta \overline{G} = [g\overline{G}_T^0(G) + h\overline{G}_T^0(H) + \ldots - a\overline{G}_T^0(A) - b\overline{G}_T^0(B) - \ldots]$$
$$+ gR^*T \ln p_G + hR^*T \ln p_H + \ldots - aR^*T \ln p_A - bR^*T \ln p_B - \ldots$$

or,

$$\Delta \overline{G} = \Delta \overline{G}_T^0 + R^*T \ln \frac{(p_G)^g (p_H)^h \ldots}{(p_A)^a (p_B)^b \ldots} \tag{2.41}$$

From Eqs. (1.9b) and (2.41)

$$\Delta \overline{G} = \Delta \overline{G}_T^0 + R^*T \ln K_p \tag{2.42}$$

If the pressures $P_A, P_B, \ldots P_G, P_H, \ldots$ are those that exist when the reactants and products are in chemical equilibrium, then $\Delta \overline{G} = 0$, and

$$\Delta \overline{G}_T^0 = -R^*T \ln K_P \tag{2.43}$$

Equation (2.43) can be used to obtain the value of $\Delta \overline{G}_T^0$, which is the change in the Gibbs free energy for a reaction at 1 atm and temperature T, from the equilibrium constant for the reactant at pressure p and temperature T. For standard conditions (1 atm and 25.00°C), Eq. (2.43) becomes

$$\Delta \overline{G}^0 = -R^*(298.15)\ln K_P = -2478.9 \ln K_p = -5707.9 \log_{10} K_p \tag{2.44}$$

where $\Delta\overline{G}^0$ is the change in the standard Gibbs free energy and K_p the equilibrium constant for the reaction under standard conditions.

We can see from Eq. (2.43) that if $\Delta\overline{G}^0_T$ has a large negative value, K_p will be large and positive, which implies from Eqs. (1.6) and (1.9a) that at equilibrium the products will be present in high concentrations. Conversely, if $\Delta\overline{G}^0_T$ is positive then $K_p < 1$, and at equilibrium the reactants will be favored over the products.

In the case of ideal solutions, the analogous expression to Eq. (2.40) is

$$\overline{G} - \overline{G}^0_T = R^*T \ln \frac{C}{C^0_T} \qquad (2.45)$$

where C is the concentration (in any appropriate units) at temperature T, and C^0_T is a standard concentration (e.g., 1 M) at temperature T. The analogous expressions to Eqs. (2.43) and (2.44) for ideal solutions are

$$\Delta\overline{G}^0_T = -R^*T \ln K_c \qquad (2.46)$$

and,

$$\Delta\overline{G}^0 = -R^*(298.15)\ln K_c = -2478.9\ln K_c = -5707.9\log_{10}K_c \quad (2.47)$$

where K_c is the equilibrium constant for the reaction, which is given by Eq. (1.6).

Exercise 2.7. Calculate $\Delta\overline{G}^0$ and K_p at 25°C for the forward reaction

$$NO(g) + O_3(g) \rightleftharpoons NO_2(g) + O_2(g)$$

Are the reactants or products favored for the forward reaction at equilibrium?

Solution. From Eq. (2.34)

$$\Delta\overline{G}^0 = \Delta\overline{G}^0_f[NO_2(g)] + \Delta\overline{G}^0_f[O_2(g)] - \Delta\overline{G}^0_f[NO(g)] - \Delta\overline{G}^0_f[O_3(g)]$$

From Appendix V we see that the values of $\Delta\overline{G}^0_f$ at 25°C for NO(g), $O_3(g)$, and $NO_2(g)$ are 86.6, 163.2, and 51.8 kJ mol^{-1}, respectively, and for $O_2(g)$ the value of $\Delta\overline{G}^0_f$ is zero. Therefore, at 25°C

$$\Delta\overline{G}^0 = (51.8 + 0 - 86.6 - 163.2) \text{ kJ} = -198 \text{ kJ}$$

or

$$\Delta\overline{G}^0 = -198 \times 10^3 \text{ J}$$

Substituting this value for $\Delta\overline{G}^0$ into Eq. (2.44) we obtain

$$\ln K_p = \frac{198 \times 10^3}{2478.9} = 79.9$$

or,

$$K_p = 5.01 \times 10^{34}$$

Since K_p is so large, the products of the forward reaction are certainly favored under equilibrium conditions at a temperature of 25°C and 1 atm.

2.8 Chemical potential; homogeneous nucleation of water-vapor condensation

If a single molecule is removed from a material in a certain phase, with temperature and pressure remaining constant, the resulting change in the Gibbs free energy of the material is called the *chemical potential* (μ) of that phase. In other words, *the chemical potential is the Gibbs free energy per molecule at constant temperature and pressure.*

Exercise 2.8. Show that when a plane surface of a liquid is in equilibrium with its vapor, the chemical potentials in the liquid and vapor phases are the same.

Solution. If a liquid is in equilibrium with its vapor, molecules may evaporate and condense without the temperature or pressure of the system changing. Hence, from Eq. (2.35), $dg = 0$. But, if g does not change when a molecule passes from the liquid to the vapor phase (or vice versa), the Gibbs free energy per molecule (i.e., the chemical potential) must be the same in the liquid phase as in the vapor phase. The pressure exerted by the vapor under these equilibrium conditions is called the *saturation vapor pressure;* it depends only on the substance being considered and its temperature.

We will now derive an expression for the difference in the chemical potentials of a vapor and its liquid at an arbitrary partial pressure e and temperature T. Applying Eq. (2.40) to one molecule of the vapor, we obtain

$$\mu_v - \mu_v^0 = kT \ln e \qquad (2.48)$$

where μ_v^0 is the chemical potential of the vapor molecule at 1 atm, and k is the Boltzmann constant. Similarly, the chemical potential (μ_{vsat}) for one molecule of the vapor at its saturation vapor pressure e_s at temperature T is given by

$$\mu_{\text{vsat}} - \mu_v^0 = kT \ln e_s \qquad (2.49)$$

Also from Exercise 2.8,

$$\mu_{\text{vsat}} = \mu_\ell \qquad (2.50)$$

where μ_ℓ is the chemical potential of a molecule in the liquid phase that is at equilibrium with the vapor phase at temperature T. From Eqs. (2.48) and (2.49)

$$\mu_v - \mu_{\text{vsat}} = kT \ln \frac{e}{e_s}$$

or using Eq. (2.50),

$$\mu_v - \mu_\ell = kT \ln \frac{e}{e_s} \qquad (2.51)$$

Let us consider now the formation of a pure droplet by condensation from its supersaturated vapor. In this process, which is referred to as *homogeneous nucleation*, the first stage in the growth process is the chance collisions of a number of molecules in the vapor phase to form a small embryonic droplet large enough to remain intact. Let V be the volume and A the surface area of such an embryonic droplet that has formed at constant temperature and pressure. If μ_ℓ and μ_v are the chemical potentials in the liquid and vapor phases, and n is the number of molecules per unit volume of liquid, the decrease in the Gibbs free energy of the system due to the condensation is $nV(\mu_v - \mu_\ell)$. Now, quite apart from any work associated with the change in volume of the system, work is done in creating the surface area of the droplet. This work may be written as $A\sigma$, where σ is the work required to create a unit area of vapor-liquid interface (called the *interfacial energy* between the vapor and the liquid, or the *surface energy* of the liquid). If this were an equilibrium transformation (which it is not) $A\sigma$ would be equal to $nV(\mu_v - \mu_\ell)$ (see Exercise 2.18). Instead, the change in the Gibbs free energy will, in general, differ from the work term $A\sigma$. Let us write

$$\Delta E = A\sigma - nV(\mu_v - \mu_\ell) \qquad (2.52)$$

then, ΔE is the net increase in the energy of the system due to the formation of the drop. Combining Eqs. (2.51) and (2.52) we obtain

$$\Delta E = A\sigma - nVkT \ln \left(\frac{e}{e_s}\right) \qquad (2.53)$$

where e and T are the vapor pressure and temperature of the system, respectively, and e_s is the saturation vapor pressure. For a droplet of radius R, Eq. (2.53) becomes

$$\Delta E = 4\pi R^2 \sigma - n \frac{4}{3} \pi R^3 kT \ln\left(\frac{e}{e_s}\right) \tag{2.54}$$

In subsaturated air $e < e_s$; therefore, $\ln(e/e_s)$ is negative and ΔE is always positive and increases with increasing R (Fig. 2.3). In other words, the larger the embryonic droplet that forms in a subsaturated vapor the greater the increase in the energy of the system. Since a system approaches an equilibrium state by reducing its energy, the formation of droplets is clearly not favored under subsaturated conditions. Even so, due to random collisions of vapor molecules, very small embryonic droplets continually form (and evaporate) in a subsaturated vapor, but they are neither numerous nor large enough to become visible as a cloud of droplets.

When air is supersaturated, $e > e_s$ and $\ln(e/e_s)$ is positive, so that ΔE in Eq. (2.54) can be either positive or negative depending upon the value of R. The variation of ΔE with R for this case is also shown in Figure 2.3, where it can be seen that ΔE initially increases with increasing R, reaches a maximum value when $R = r$, and then decreases with increasing R. Hence, in a supersaturated vapor, embryonic droplets with $R < r$ tend to evaporate, but droplets that manage to grow by

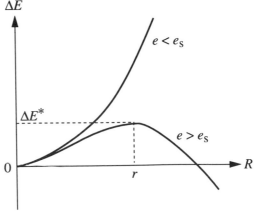

Figure 2.3. Increase ΔE in the energy of a system due to the formation of a droplet of radius R from water vapor with pressure e; e_s is the saturation vapor pressure with respect to a plane surface of the liquid at the temperature of the system.

chance collisions to a radius that just exceeds r will continue to grow by condensation from the vapor phase, because, by so doing, they cause a decrease in the total energy of the system.

In the region just around $R = r$, a droplet can grow or evaporate infinitesimally without any change in the energy of the system. We can obtain an expression for r in terms of e by setting $\partial(\Delta E)/\partial R = 0$ at $R = r$. Hence, from Eq. (2.54)

$$r = \frac{2\sigma}{nkT \ln(e/e_s)} \tag{2.55}$$

Equation (2.55) is referred to as *Kelvin's formula*. It is important, for example, in understanding the condensation and evaporation of small water drops in air, such as those in clouds.

Exercise 2.9. What is the relative humidity at 10°C just above the surface of a water droplet with a radius of 0.010 μm? If the droplet were placed in an enclosure having this same relative humidity, would the droplet be in stable or unstable equilibrium? (The surface energy of water at 10°C is 0.076 J m^{-2}, and the number density of molecules in water is 3.3×10^{28} m^{-3}.)

Solution. Kelvin's formula [Eq. (2.55)] allows us to determine the radius of a droplet that is in equilibrium (i.e., neither evaporating nor condensing) with a water vapor pressure e. However, if a droplet neither evaporates nor grows when the air surrounding it has vapor pressure e, it is because the vapor pressure just above the surface of the droplet is equal to e. Thus, we can interpret Eq. (2.55) as giving the vapor pressure e (or relative humidity RH = 100 e/e_s) just above the surface of a droplet of radius r. Hence, from Eq. (2.55)

$$\text{RH} = 100 \frac{e}{e_s} = 100 \exp\left(\frac{2\sigma}{nkTr}\right) \tag{2.56}$$

Substituting, $\sigma = 0.076$ J m^{-2}, $n = 3.3 \times 10^{28}$ m^{-3}, $k = 1.381 \times 10^{-23}$ J deg^{-1} molecule^{-1}, $T = 283$K, and $r = 0.010$ μm $= 1.0 \times 10^{-8}$ m into Eq. (2.56) gives RH $\approx$ 110%. (*Note:* Substitution of the numerical values given in this problem into Eq. (2.56) gives RH = 113%. However, the numerical values are given to an accuracy of just two significant figures. Therefore, the derived value of RH is also accurate to just two significant figures; hence 113% must be rounded to 110%.)

If the droplet is placed in an enclosure with a relative humidity of 110%, it will be in equilibrium. However, if the droplet should evaporate very slightly, the relative humidity just above its surface would increase above 110% [see Eq. (2.56)]. This would cause water vapor to diffuse away from just above the surface of the droplet to the drier (110% relative humidity) air in the enclosure. The droplet would therefore continue to evaporate in order to maintain the relative humidity just above its surface. A similar argument will show that if the droplet should grow very slightly, it would continue to grow. Hence, a droplet with radius 0.010 μm is in *unstable* equilibrium in an environment with a relative humidity 110%. This is an example of a physical equilibrium that is controlled by chemical potentials.

Exercises

2.10.   Answer, interpret, or explain the following in light of the principles discussed in this chapter.

(a) An equilibrium chemical state represents a compromise between the drive for molecules to assume a state of minimum energy and the drive toward a state of maximum entropy (i.e., maximum molecular chaos). Consider, for example, the reaction $H_2(g) \rightleftarrows 2H(g)$.

(b) The development of life (a relatively ordered state) on Earth does not contradict the second law of thermodynamics.

(c) The work done by a system in a reversible process is greater than the work done by the system in an irreversible process between the same two states.

(d) The second law of thermodynamics is sometimes expressed in the form: "The entropy of the universe is continually increasing."

(e) Water can freeze spontaneously below 0°C even though this produces a decrease in the entropy of the system.

(f) Some exothermic chemical reactions do not occur spontaneously.

(g) List some spontaneous processes that are not exothermic.

(h) All chemical compounds will dissociate into the elements at sufficiently high temperatures. (*Hint:* Consider the Gibbs–Helmholtz equation.)

(i) Neither the heat absorbed (or released) nor the work

done by (or on) a system is a function of state, but the difference between these two quantities, namely, the change in the internal energy of a system, is a function of state.

(j) When a system undergoes a transformation, the change in its entropy can be computed from Eq. (2.17) by taking the system from its initial to its final state by a *reversible* path. However, entropy is a state function and therefore does not depend on the path taken. Explain this apparent paradox.

(k) By measuring the equilibrium radius (*r*) of a droplet in a given environment, the relative humidity (RH) of the environment could be derived from Eq. (2.56). What would be the practical problems in utilizing this technique as a sensitive measure of RH?

2.11. Use Eq. (2.13) to show that:

$$\log_{10}\frac{K_2}{K_1}=\frac{\Delta \overline{H}^0_{rx}}{2.303R^*}\frac{T_2-T_1}{T_1T_2}$$

where K_1 and K_2 are the equilibrium constants at temperatures T_1 and T_2, respectively, and $\Delta \overline{H}^0_{rx}$ is the molar standard enthalpy (heat) of reaction. Assuming that $\Delta \overline{H}^0_{rx}$ is independent of temperature, show that the predictions of this relationship with respect to the variations of K_p with temperature for exothermic and endothermic reactions are consistent with deductions based on LeChatelier's principle.

2.12. If the equilibrium constant for a chemical reaction is 2.20×10^{-2} at 0°C and the molar standard enthalpy (heat) of reaction is -3.01×10^5 J mol^{-1}, what is the value of the equilibrium constant for the reaction at 20°C?

2.13. Calculate the molar standard enthalpies (or heats) of reaction $\Delta \overline{H}^0_{rx}$ for the reactions:

(a) $N_2(g)+O_2(g) \rightarrow 2NO(g)$
(b) $O_3(g)+NO(g) \rightarrow NO_2(g)+O_2(g)$
(c) $2SO_2(g)+O_2(g) \rightarrow 2SO_3(g)$

Are the reactions endothermic or exothermic at 1 atm and 25°C?

2.14.   Determine the molar standard enthalpy (or heat) of reaction for:

$$C(s) + 2H_2(g) \rightleftarrows CH_4(g)$$

given that the values of $\Delta \overline{H}_{rx}^0$ for the reactions:

$$C(s) + O_2(g) \rightleftarrows CO_2(g)$$
$$H_2(g) + \tfrac{1}{2}O_2(g) \rightleftarrows H_2O(l)$$

and,

$$CH_4(g) + 2O_2(g) \rightleftarrows CO_2(g) + 2H_2O(l)$$

are -393.5, -285.6, and -889.9 kJ mol^{-1}, respectively.

2.15.   Assuming the truth of the second law of thermodynamics, as expressed by Eq. (2.20), prove that heat will not flow unaided from a cold to a hot body in an isolated system.

2.16.   State whether the following reactions are likely to result in an increase or decrease in the entropy of the system:
(a) $3H_2(g) + N_2(g) \rightarrow 2NH_3(g)$
(b) $2SO_2(g) + O_2(g) \rightarrow 2SO_3(g)$
(c) $2C_2H_6(g) + 7O_2(g) \rightarrow 4CO_2(g) + 6H_2O(l)$

2.17.   Use the Gibbs–Helmholtz equation in the form of Eq. (2.28) to determine the effect of temperature (T) on the spontaneity of chemical reactions for which (a) dh is negative and ds is positive, (b) dh is positive and ds is negative, (c) dh and ds are both negative, and (d) dh and ds are both positive.

2.18.   In formulating the first law of thermodynamics in Section 2.1 we tacitly assumed that the only external work that a system can do is the work of expansion ($p\,d\alpha$). However, a system may also perform external work by other means (e.g., chemical or electrical). Therefore, in general, the first law of thermodynamics for a unit mass of a system should be written as

$$dq = du + dw_{total}$$

where dw_{total} is the total external work done by the system. If da is the external work done by a unit mass of a system *over and above* any $p\,d\alpha$ work, that is

$$da = dw_{total} - p\,d\alpha,$$

show that for a reversible transformation at constant pressure and temperature, da is equal to the decrease in the Gibbs free energy ($da = -dg$).

2.19. The *Helmholtz free energy* of a unit mass of a system is defined as

$$f = u - Ts$$

Show that the criterion for the thermodynamic equilibrium of a system at constant temperature and *volume* is that the Helmholtz free energy of the system have a minimum value.

2.20. Using the definition of f given in Exercise 2.19 and the definition of dw_{total} given in Exercise 2.18, show that for a reversible, isothermal transformation

$$dw_{total} = -df$$

2.21. Calculate $\Delta \overline{G}_f^0$ and K_p at 298K for the forward reaction

$$2NO(g) + O_2(g) \rightleftarrows 2NO_2(g)$$

Are the reactants or products favored for the forward reaction at equilibrium?

2.22. By considering the signs of $\Delta \overline{S}^0$ and $\Delta \overline{H}_{rx}^0$ determine whether the reaction

$$2NO(g) \rightleftarrows N_2(g) + O_2(g)$$

is favored at high or low temperatures. Consider the implication of your conclusion with respect to NO(g) pollution from combustion processes.

2.23. Derive the relationship given in Exercise 2.11 using the Gibbs–Helmholtz equation (2.29) and Eq. (2.43).

2.24. Calculate the equilibrium constants K_p at 298K and 500K for the reaction

$$H_2O_2(g) \rightarrow H_2O(g) + \tfrac{1}{2}O_2(g)$$

2.25. Calculate the change in chemical potential when an ideal gas expands from 1 atm to 0.5 atm at 15°C.

2.26. Show that

$$dq = dh - \alpha \, dp$$

where the symbols are as defined in Section 2.1.

2.27.   Show that if a chemical reaction occurs at pressures near 1 atm and involves only solids and/or liquids

$$dH = dU$$

where dH and dU are the changes in enthalpy and internal energy of the system during the chemical reaction. If, on the other hand, gases are involved in the chemical reaction, show that

$$dH = dU + R^*T dn + R^*n dT$$

where dn is the change in the number of moles of gas during the chemical reactions and dT the change in temperature.

2.28.   Show that for an ideal gas

$$\frac{dq_{rev}}{T} = c_p \frac{dT}{T} - R \frac{dp}{p}$$

where the symbols have the same meanings as those defined in Section 2.3.

2.29.   Electrochemical cells (i.e., "wet" batteries) are powered by chemical reactions. Derive an expression for the number of moles (n) of electrons that are transferred through a cell in terms of the change in the Gibbs free energy (ΔG) of the system due to the chemical reaction, the difference in volts (ΔE) across the terminals of the battery, and the charge in coulombs (F) on one mole of electrons. (*Hint:* Use the result from Exercise 2.18. Also, recall that when a charge of 1 coulomb passes through a potential difference of 1 volt, 1 J of work is involved.)

Notes

1 If the internal energy of a gas depends on its temperature only (and not its volume), it is said to obey *Joule's law*. One of the requirements of an ideal gas is that it obey Joule's law.

2 Note that the standard temperature is defined as 0°C (see Section 1.1), but *standard states* are defined at a temperature of 25°C (or 298 K).

3 Although one atmosphere (1 atm) is not the SI unit of pressure, it does not matter here because the pressures in Eq. (2.39) occur as a ratio (dp/p).

3
Chemical kinetics

Our previous discussion of chemical equilibria and chemical thermo-dynamics allows us to assess whether or not a chemical reaction will proceed in a certain direction, and what the concentrations of the reactants and products will be when a system is in chemical equilib-rium. In this chapter we are concerned with how *fast* reactants are converted into products, some of the factors upon which the rate of conversion depends, and the sequence of steps by which the conver-sion occurs. These subjects are the province of *chemical kinetics*.

3.1 Reaction rates

The rate of a chemical reaction could be measured by the rate at which the concentration of one of the reactants decreases or one of the products increases with time. In this case, however, as the following exercise demonstrates, the rate would depend on which reagent was considered.

Exercise 3.1. Compare the rate of disappearance of $N_2O_5(g)$ and the rates of formation of $NO_2(g)$ and $O_2(g)$ in the reaction

$$2N_2O_5(g) \rightarrow 4NO_2(g) + O_2(g)$$

(A single arrow from left to right indicates that we need only be concerned with the forward reaction.)

Solution. For every 2 moles of $N_2O_5(g)$ that disappear, 4 moles of $NO_2(g)$ and 1 mole of $O_2(g)$ are formed. Hence,

$$-\frac{1}{2}\left(\frac{d[N_2O_5(g)]}{dt}\right) = \frac{1}{4}\left(\frac{d[NO_2(g)]}{dt}\right) = \left(\frac{d[O_2(g)]}{dt}\right)$$

It is clear from Exercise 3.1 that for the general chemical reaction

$$aA + bB + \ldots \rightarrow gG + hH \ldots \tag{3.1}$$

we have

$$-\frac{1}{a}\frac{d[A]}{dt} = -\frac{1}{b}\frac{d[B]}{dt} = \ldots = \frac{1}{g}\frac{d[G]}{dt} = \frac{1}{h}\frac{d[H]}{dt} = \ldots \tag{3.2}$$

where [A], [B], ... are the molar concentrations. Hence, any one of the quantities in Eq. (3.2) is referred to as the *reaction rate* for Reaction (3.1). The reaction rate is usually expressed in units of mol L^{-1} s^{-1} (or M s^{-1}, where M is molarity).

Many reactions occur at a decreasing rate with increasing time. This is because the reaction rate diminishes as the concentrations of the reactants diminish. For Reaction (3.1), the reaction rate can often be expressed as

$$\text{Reaction rate} = k[A]^m[B]^n \ldots \tag{3.3}$$

Equation (3.3) is called the *rate law* or *rate equation* for Reaction (3.1). The exponents m and n are generally integers or half-integers; m is called the *order* of the reaction with respect to A, and n the order with respect to B, etc. The *overall order* of the reaction is $m + n + \ldots$. The orders m, n, ... must be determined experimentally, because, in general, they cannot be predicted theoretically or deduced from Reaction (3.1).

The term k in Eq. (3.3) is called the *rate coefficient* (or, more formally, the *specific reaction rate coefficient*, since it is numerically equal to the rate of reaction if all concentrations were unity). Each reaction is characterized by a value of k at each temperature. The units of k depend on the overall order of the reaction.

One method for determining the exponents m, n, ... in Eq. (3.3) is from the *initial* reaction rates for several different sets of initial concentrations of A, B, ... For example, if when [A] is doubled, and [B] ... are held constant, the initial reaction rate doubles, $m = 1$; if it quadruples, $m = 2$, etc. (See Exercise 3.8.)

Consider a reaction that is first order in just one reactant A and for which $a = 1$ in Reaction (3.1). Then,

$$-\frac{d[A]}{dt} = k[A] \tag{3.4}$$

Therefore,

$$\int_{[A]_0}^{[A]_t} \frac{d[A]}{[A]} = -k \int_0^t dt$$

where $[A]_0$ and $[A]_t$ are the initial concentration of A and the concentration at time t, respectively. Hence,

$$\ln \frac{[A]_t}{[A]_0} = -kt$$

or, converting to logarithms to the base 10 (indicated by "log"),

$$\log[A]_t = \frac{-kt}{2.303} + \log[A]_0 \tag{3.5}$$

It follows from Eq. (3.5) that for a first-order reaction in A, a plot of log $[A_t]$ versus t will give a straight line of slope $-k/2.303$ and intercept on the ordinate of log $[A]_0$.

For a reaction that is second order (*bimolecular*) in just one reactant A, and for which $a = 1$ in Reaction (3.1),

$$-\frac{d[A]}{dt} = k''[A]^2 \tag{3.6}$$

where k'' is the rate coefficient. Therefore,

$$\int_{[A]_0}^{[A]_t} \frac{d[A]}{[A]^2} = -k'' \int_0^t dt$$

and,

$$\frac{1}{[A]_t} = k''t + \frac{1}{[A]_0} \tag{3.7}$$

Hence, in this case, a plot of $\frac{1}{[A]_t}$ versus t will give a straight line of slope k'' and intercept $\frac{1}{[A]_0}$.

In some cases it is possible and convenient to define a *pseudo first-order rate coefficient* even though a reaction is of higher order. Consider, for example,

$$OH(g) + SO_2(g) + M \rightarrow HOSO_2(g) + M$$

which is a (*termolecular*) third-order reaction[1] that is of importance in the atmosphere. The rate law for this reaction is

$$-\frac{d[OH(g)]}{dt} = k'''[OH(g)][SO_2(g)][M]$$

In the atmosphere [M] and $[SO_2(g)]$ are generally essentially constant; therefore, their concentrations can be combined with the value of k''' to give

$$-\frac{d[OH(g)]}{dt} = k_{pseudo}[OH(g)]$$

where k_{pseudo} is a pseudo first-order rate coefficient.

3.2 Reaction mechanisms

How are reactants converted into products in a chemical reaction? The simplest process is *single step and unimolecular,* in which one molecule gains sufficient energy to break a chemical bond. For example,

$$O_3^*(g) \rightarrow O_2(g) + O(g)$$

where the superscript asterisk indicates that the ozone molecule is sufficiently energized (or excited) to dissociate. Single-step (or *elementary*) processes can also involve two reactant molecules (a *bimolecular process*). For example,

$$NO(g) + O_3(g) \rightarrow NO_2(g) + O_2(g)$$

Or, three reactant molecules (a *termolecular* process) can be involved

$$O(g) + O_2(g) + N_2(g) \rightarrow O_3(g) + N_2(g)$$

In this example, as in most termolecular processes, two molecules combine, and the third molecule removes the excess energy produced by the combination. (If this excess energy is not removed, the O_3 molecule is likely to dissociate again into O and O_2.) Elementary processes involving more than three molecules are not known, since the chance that more than three molecules will collide simultaneously is so small that such a process would produce a negligibly small reaction rate.

Chemical reactions generally involve several single steps or elementary processes, which together form what is called the *reaction mechanism*. For example, the reaction

$$NO_2(g) + CO(g) \rightarrow NO(g) + CO_2(g)$$

occurs through the following two elementary processes, each of which is bimolecular

$$NO_2(g) + NO_2(g) \rightarrow NO_3(g) + NO(g)$$
$$NO_3(g) + CO(g) \rightarrow NO_2(g) + CO_2(g)$$

In a multistep chemical reaction, the elementary processes must add to give the overall chemical reaction. Thus, adding the above two elementary processes,

$$NO_2(g) + NO_2(g) + NO_3(g) + CO(g) \rightarrow$$
$$NO_3(g) + NO(g) + NO_2(g) + CO_2(g)$$

and, by canceling species that appear on both sides of the reaction, we obtain

$$NO_2(g) + CO(g) \rightarrow NO(g) + CO_2(g)$$

Although $NO_3(g)$ appears in the elementary processes, it is neither a reactant nor a product in the overall chemical reaction; it is formed in one elementary process and consumed in another. Such a reagent is called an *intermediate*.

Although the order of a chemical reaction cannot be predicted from the overall reaction, the order of an elementary process is predictable. For example, for the general unimolecular process

$$A \rightarrow \text{products}$$

the rate of decomposition of A at any time will be proportional to [A], therefore,

$$-\frac{d[A]}{dt} = k[A]$$

For the general bimolecular reaction

$$A + B \rightarrow \text{products}$$

the rate of collision between the A molecules and the B molecules will be proportional to [A] and [B], therefore,

$$-\frac{d[A]}{dt} = k[A][B]$$

For the special case of a bimolecular reaction

$$A + A \rightarrow products$$

that is,

$$2A \rightarrow products$$

$$-\frac{d[A]}{dt} = k[A]^2$$

We conclude that the *order for each reactant in a single-step (elementary) process is equal to the coefficient of that reactant in the chemical equation for that process, and, for an elementary process, the overall order is the same as the molecularity* (i.e., a unimolecular process is first order, a bimolecular process is second order, etc.) The converse does *not* hold; that is, not all first-order chemical reactions are unimolecular elementary processes, etc.

How is the order and rate of an overall chemical reaction related to the orders and rates of the elementary processes that comprise the reaction? The answer is simple for most reactions. Since the overall reaction can be no faster than its slowest step (called the *rate-determining step*), the rate law for the overall reaction is closely related to the rate law for this step.

A common method for determining a reaction mechanism is first to determine the rate law experimentally, and then to postulate one or more elementary processes that are consistent with the overall rate law. The following two problems illustrate the general approach.

Exercise 3.2. The following mechanism has been proposed for the formation of $N_2O_5(g)$ from $NO_2(g)$ and $O_3(g)$ in the gas phase within clouds

$$NO_2(g) + O_3(g) \rightarrow NO_3(g) + O_2(g) \qquad \qquad (i)$$

$$NO_3(g) + NO_2(g) + M \rightarrow N_2O_5(g) + M \qquad \qquad (ii)$$

(a) Write down the overall chemical reaction. (b) What is the intermediate? (c) What is the rate law for each step? (d) If the experimentally determined rate law for the overall chemical reaction is

$$-\frac{d[NO_2(g)]}{dt} = k[NO_2(g)][O_3(g)]$$

what can be concluded about the relative speeds of steps (i) and (ii)?

Solution. (a) The overall chemical reaction is obtained by adding the elementary processes and canceling the species that appear on both sides of the equation. Hence, adding Reactions (i) and (ii), we get

$$NO_2(g) + O_3(g) + NO_3(g) + NO_2(g) + M \rightarrow NO_3(g) + O_2(g) + N_2O_5(g) + M$$

or,

$$2NO_2(g) + O_3(g) \rightarrow N_2O_5(g) + O_2(g)$$

(b) Since $NO_3(g)$ is formed in Reaction (i) and consumed in Reaction (ii), it is an intermediate.

(c) Since Reaction (i) is a bimolecular process, its rate law is second order and is given by

$$-\frac{d[NO_2(g)]}{dt} = k_1[NO_2(g)][O_3(g)]$$

Similarly, for Reaction (ii)

$$-\frac{d[NO_3(g)]}{dt} = k_2[NO_3(g)][NO_2(g)]$$

(d) Comparing the rate law for the overall reaction

$$-\frac{d[NO_2(g)]}{dt} = k[NO_2(g)][O_3(g)]$$

with the rate laws given in (c) for the two elementary processes, we see that it is identical to the rate law for Reaction (i). Hence, it is this process that is rate controlling. Therefore, Reaction (i) proceeds at a slower speed than Reaction (ii).

Exercise 3.3. The reaction mechanism for

$$H_2(g) + Br_2(g) \rightarrow 2HBr(g)$$

which is important in volcanic plumes is

$$Br_2(g) + M \xrightarrow{k_{1f}} 2Br(g) + M \qquad \text{(ia)}$$
$$2Br(g) + M \xrightarrow{k_{1r}} Br_2(g) + M \qquad \text{(ib)}$$

$\left.\begin{array}{c} \\ \\ \end{array}\right\}$ (fast equilibrium)

$$\mathrm{Br(g) + H_2(g) \overset{k_2}{\rightarrow} HBr(g) + H(g)} \quad \text{(ii)} \quad \text{(slow)}$$

$$\mathrm{H(g) + Br_2(g) \overset{k_3}{\rightarrow} HBr(g) + Br(g)} \quad \text{(iii)} \quad \text{(fast)}$$

where the k's are the rate coefficients for each elementary process, and k_{1f} and k_{1r} are the forward and reverse rate coefficients for (i). Derive an expression for $\dfrac{d[\mathrm{HBr(g)}]}{dt}$ in terms of $[\mathrm{H_2(g)}]$ and $[\mathrm{Br_2(g)}]$.

Solution. Since (ii) is the slowest reaction, it is the rate-determining step. Therefore, from Reaction (ii)

$$\frac{d[\mathrm{HBr(g)}]}{dt} = k_2[\mathrm{H_2(g)}][\mathrm{Br(g)}]$$

Also, for

$$\mathrm{Br_2(g) \rightleftarrows 2Br(g)}$$

$$K_c = \frac{[\mathrm{Br(g)}]^2}{[\mathrm{Br_2(g)}]}$$

where, K_c is the equilibrium constant. Hence,

$$\frac{d[\mathrm{HBr(g)}]}{dt} = k_2 K_c^{1/2}[\mathrm{H_2(g)}][\mathrm{Br_2(g)}]^{1/2} = k[\mathrm{H_2(g)}][\mathrm{Br_2(g)}]^{1/2}$$

This problem illustrates how half-integer rate laws arise. Also, it shows how the rate coefficient k for a reaction can be related to other parameters (k_2 and K_c in this case).

Just because a series of elementary processes is consistent with a chemical reaction and its experimentally determined rate law, these elementary processes are not necessarily responsible for the reaction. This is because more than one series of steps can satisfy a reaction and a rate law (see Exercise 3.13). In general, information in addition to the rate law is required if a reaction mechanism is to be determined.

3.3 Reaction rates and equilibria

When a chemical reaction is in equilibrium, the forward and reverse reactions occur at the same rate. Using this principle we can derive a general relationship between the equilibrium constant and the forward and reverse rate constants for any *elementary process*. Let

$$A + B \xrightarrow{k_f} AB$$

be an elementary process. Then, the rate of increase of [AB] is $k_f[A][B]$. The reverse elementary process is

$$AB \xrightarrow{k_r} A + B$$

Hence, the rate of decrease of [AB] is $k_r[AB]$. At equilibrium there is no net change of [AB]. Therefore,

$$k_f[A][B] = k_r[AB]$$

or,

$$\frac{k_f}{k_r} = \frac{[AB]}{[A][B]}$$

But the right-hand side of the last relation is the equilibrium constant K_c for the forward reaction. Therefore,

$$\frac{k_f}{k_r} = K_c \tag{3.8}$$

which is the required relationship for an elementary process.

In the case of a chemical reaction that proceeds by a multistep process, each elementary process must be in equilibrium if Eq. (3.8) is to be applied to each step. This is called the *principle of detailed balancing;* it permits a relationship to be derived between the rate coefficients for the elementary processes and the equilibrium constant for the overall reaction.[2]

Exercise 3.4. If the reaction mechanism for

$$2NO_2(g) + O_3(g) \rightarrow N_2O_5(g) + O_2(g)$$

is

$$NO_2(g) + O_3(g) \rightarrow NO_3(g) + O_2(g) \tag{i}$$
$$NO_3(g) + NO_2(g) + M \rightarrow N_2O_5(g) + M \tag{ii}$$

derive an expression for the equilibrium constant K_c for the forward reaction of the overall equation in terms of the forward and reverse rate coefficients for the elementary processes.

Solution. The condition for equilibrium is that each elementary process be in equilibrium. Hence, if k_{1f} and k_{1r} are the forward and reverse rate coefficients for (i) and k_{2f} and k_{2r} are the corresponding quantities for (ii), we have

$$k_{1f}[NO_2(g)][O_3(g)] = k_{1r}[NO_3(g)][O_2(g)]$$

and,

$$k_{2f}[NO_3(g)][NO_2(g)] = k_{2r}[N_2O_5(g)]$$

Solving each of these equations for the intermediate $NO_3(g)$, equating the results and rearranging leads to

$$\frac{k_{1f}\,k_{2f}}{k_{1r}\,k_{2r}} = \frac{[N_2O_5(g)][O_2(g)]}{[NO_2(g)]^2[O_3(g)]}$$

But, the right-hand side of the last expression is the equilibrium constant K_c for the forward reaction of the overall equation. Hence,

$$K_c = \frac{k_{1f}k_{2f}}{k_{1r}k_{2r}}$$

3.4 Collision theory of gaseous reactions

In this section we consider the factors that determine the magnitude of the rate coefficient for a gaseous bimolecular reaction.

If the effective circular cross section of the molecule for collisions has diameter ρ, two molecules will collide if their centers come within a distance ρ of each other. If we now imagine that all of the molecules except one (which we will call X) shrink to points, X will still collide with the other molecules when they come within a distance ρ of each other provided that we artificially expand the diameter of X to 2ρ. Now, in unit time the expanded molecule X, which has an artificial radius ρ, will sweep out a volume $\pi\rho^2\bar{c}$, where $\bar{c}$ is the average speed of a molecule. Therefore, if there are n "point" molecules per unit volume, and we assume that all of these molecules are stationary, the number of molecules with which X will collide in unit time will be $\pi\rho^2\bar{c}n$.[3]

If we now consider molecules A (in concentration n_A) colliding with molecules B (in concentration n_B), the number of collisions per second that *one* A molecule makes with the B molecules is $\pi\rho^2\bar{c}n_B$. This expression gives the maximum chemical reaction rate, assuming that each collision between A and B molecules results in a reaction.

Exercise 3.5. Calculate the approximate maximum rate for a gaseous bimolecular chemical reaction at 1 atm and 0°C, given that $\rho = 3 \times 10^{-10}$ m and $\bar{c} = 5 \times 10^2$ m s^{-1}. How long would it take to consume the reacting gases at this rate?

Solution. We must first determine the number of molecules per unit volume (n_o) of each gas at 1 atm and 0°C. This can be found using the ideal gas equation in the form of Eq. (1.8g)

$$p = n_o kT$$

where k is the Boltzmann constant. Substituting $p = 1$ atm $= 1.01 \times 10^5$ Pa, $T = 273$K and $k = 1.38 \times 10^{-23}$ J deg^{-1} molecule^{-1} into this equation, we obtain $n_o = 2.68 \times 10^{25}$ m^{-3}. Hence, the total rate of collision (i.e., the maximum chemical reaction rate) is equal to

$$\pi \rho^2 \bar{c} n_A n_B = (3.14)(3 \times 10^{-10})^2 (5 \times 10^2)(2.68 \times 10^{25})^2 \text{ molecules m}^{-3} \text{ s}^{-1}$$
$$= 1 \times 10^{35} \text{ molecules m}^{-3} \text{ s}^{-1}$$

Or, since there are Avogadro's number ($= 6.022 \times 10^{23}$) of molecules in 1 mole, and 1 L $= 10^{-3}$ m^3, the maximum chemical reaction rate is

$$\frac{1 \times 10^{35} \times 10^{-3}}{6.022 \times 10^{23}} = 2 \times 10^8 \text{ mol liter}^{-1} \text{ s}^{-1}$$

The time required to consume one of the gases, consisting of 2.68×10^{25} molecules m^{-3}, at a rate of 1×10^{35} molecules m^{-3} s^{-1} would be

$$\frac{2.68 \times 10^{25}}{1 \times 10^{35}} \simeq 3 \times 10^{-10} \text{ s}$$

Although a few chemical reactions actually proceed at the enormous rate calculated in Exercise 3.5, most reactions occur at a much slower rate. Hence, factors other than the mere collisions of molecules must be involved in chemical reactions. Since the rate of many chemical reactions varies markedly with temperature, we must now consider how temperature may affect a chemical reaction.

Shown in Figure 3.1 is the distribution of the kinetic energies of the molecules in a gas for two temperatures. The number of molecules with kinetic energy $\geq E_a$ is proportional to the shaded area in Figure 3.1. It can be seen that if E_a is fairly large, the number of molecules with energy $\geq E_a$ is very sensitive to temperature. Hence, if a certain minimum value of E_a is required for two colliding molecules to react chemically, it is apparent why the chemical reaction rate should be both smaller and more temperature sensitive than the collision rate.

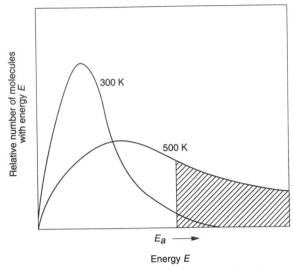

Figure 3.1. Distribution of the kinetic energies of the molecules in a gas plotted for two temperatures.

The minimum value of the relative kinetic energy of molecular motion that is required for a chemical reaction to occur is called the *activation energy* (E_a).[4] Differences in activation energies are largely responsible for the large range in the magnitudes of chemical reaction rates.

Another important factor that determines a chemical reaction rate is the orientation of the colliding molecules. The probability of a particular collision being favorable to a reaction is generally much less than unity.

Further insights into the concept of an activation energy is provided by the theory of transition states. In this theory it is hypothesized that an intermediate species, called an *activation complex,* can form during a collision between molecules. This species exists very briefly, and then either dissociates back to the original molecules or forms new molecules. In an activated complex, old bonds are stretched to the breaking point and new bonds are partially formed. An activated complex can form, thereby providing a chance for a chemical reaction to occur, only if the colliding molecules possess sufficient relative kinetic energy. The required energy is the activation energy.

Consider, for example, the reaction

$$H_2(g) + I_2(g) \rightarrow 2HI(g)$$

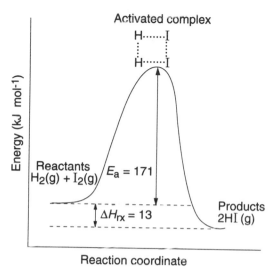

Figure 3.2. Energy profile as a function of reaction coordinate for the reaction $H_2(g) + I_2(g) \rightarrow 2HI(g)$.

This reaction may be represented as follows

$$
\begin{array}{ccccc}
\text{H} \quad \text{I} & & \text{H}\ldots\text{I} & & \text{H—I} \\
| + | & \rightarrow & & \rightarrow & | \quad | \\
\text{H} \quad \text{I} & & \text{H}\ldots\text{I} & & \text{H—I}
\end{array}
$$

reactants   activated   products
     complex

The energy profile for this reaction is shown in Figure 3.2, where the abscissa is the reaction coordinate, which represents the progress from reactants to products. The difference in energies between the reactants and products is the enthalpy of the reaction (ΔH_{rx}). The reaction is slightly exothermic, with an enthalpy change of 13 kJ mol^{-1}. However, in order for the reactants to form products, a large energy barrier (E_a) has to be overcome. This is the energy required to form the activated complex which, for the reaction considered here, is 171 kJ mol^{-1}.

3.5 The effect of temperature on reaction rates: the Arrhenius' relation

The general tendency for chemical reactions to proceed faster at higher temperatures can be understood qualitatively in terms of the concepts presented in Section 3.4. The higher the temperature, the greater will be the probability that the relative kinetic energy associated with the collision between two molecules will exceed the required activation energy, and therefore the more likely it will be for the reaction to occur.

Arrhenius found that the rate coefficient k for a chemical reaction generally varies with temperature according to the relation

$$k = A \exp\left(\frac{-E_a}{R^*T}\right) \qquad (3.9)$$

where E_a is the activation energy (units J mol^{-1}), R^* the universal gas constant, T the temperature, and A a constant (which has the same units as A) called the *frequency factor* (which is related to the frequency of molecular collisions and the probability that the collisions are favorably oriented for a chemical reaction).

From Eq. (3.9)

$$\ln k = \ln A - \frac{E_a}{R^*T}$$

or,

$$\log k = \left(\frac{-E_a}{2.303\,R^*}\right)\frac{1}{T} + \log A$$

Hence, a plot of $\log k$ versus $\frac{1}{T}$ should be a straight line of slope $\frac{-E_a}{2.303\,R}$ and the intercept at $\frac{1}{T}=0$ is $\log k = \log A$. This provides a means for determining E_a and A.

Exercise 3.6. The reaction

$$2NO_2(g) \rightarrow 2NO(g) + O_2(g)$$

has a rate coefficient of 1.0×10^{-10} s^{-1} at 300 K and an activation energy of 111 kJ mol^{-1}. What is the rate coefficient at 273 K? ($R^* = 8.314 \times 10^{-3}$ kJ mol^{-1} K^{-1}.)

Solution. Substituting $k=1.0\times10^{-10}$ s^{-1}, $E_a=111$ kJ mol^{-1}, $T=300°$K, and $R^*=8.314\times10^{-3}$ kJ mol^{-1} K^{-1} into Eq. (3.9) gives

$$A=\frac{1.0\times10^{-10}}{\exp\left(\dfrac{-111}{(8.314\times10^{-3})300}\right)}\,s^{-1}=2.1\times10^{9}s^{-1}$$

If we now assume (as is usually done) that A and E_a do not vary with temperature, we can substitute $A=2.1\times10^9$ s^{-1}, $E_a=111$ kJ mol^{-1}, $R^*=8.314\times10^{-3}$ kJ mol^{-1} K^{-1}, and $T=273$K into Eq. (3.9) to obtain the rate constant at 273K. Hence,

$$k_{273K}=2.1\times10^9\exp\left(\frac{-111}{(8.314\times10^{-3})273}\right)s^{-1}=1.2\times10^{-12}s^{-1}$$

3.6 Catalysis

A *catalyst* is a substance that increases the rate of a chemical reaction without itself undergoing a permanent chemical change. Thus, the catalyst does not appear in the net chemical equation for the reaction, although its presence may be indicated above the arrow in the equation. Generally, a catalyst increases the chemical reaction rate by providing an alternative pathway that has a lower activation energy.

If the reactants and the catalyst are in the same phase, the catalysis is said to be *homogeneous*. A *heterogeneous* catalyst exists in a different phase from the reactant molecules.[5]

To illustrate a homogeneous catalyst consider the reaction

$$2H_2O_2(aq)\rightarrow2H_2O(l)+O_2(g)$$

In the absence of a catalyst this reaction proceeds in the forward direction, but very slowly. However, it can be catalyzed by bromine (Br_2). This catalysis occurs in two steps. First

$$Br_2(aq)+H_2O_2(aq)\rightarrow2Br^-(aq)+2H^+(aq)+O_2(g)$$

then,

$$2Br^-(aq)+H_2O_2(aq)+2H^+(aq)\rightarrow Br_2(aq)+2H_2O(l)$$

Adding the last two equations we get for the net reaction

$$2H_2O_2(aq)\rightarrow2H_2O(l)+O_2(g)$$

Since bromine does not appear in the net reaction, it is a catalyst. Moreover, since it acts in the aqueous phase on aqueous-phase reactants, it is a homogeneous catalyst.

Heterogeneous catalysts often consist of finely divided metals or metal oxides. Molecules from the gaseous or liquid phase can adsorb[6] onto the solid surfaces of the metals where their reactions with each other may be enhanced. For example, the catalytic converter, which is required on all automobiles in the United States, consists of beads impregnated with cupric oxide [$CuO(s)$] and chromium oxide [$Cr_2O_3(s)$]. Their purpose is as follows. Nitrogen oxides and various unburned hydrocarbons, $C_xH_y(g)$, are emitted from cars and are involved in the formation of photochemical smog. In addition, the air pollutant carbon monoxide, $CO(g)$, is present in the emission from automobiles. The catalytic converter is designed to reduce the nitrogen oxides to nitrogen gas and to oxidize the $CO(g)$ and unburned hydrocarbons to $CO_2(g)$ and water. Reduction of the nitrogen oxides is favored thermodynamically, but the reaction is very slow; the most effective catalysts are transition-metal oxides and noble metals. Although the same types of materials also catalyze the combustion of $CO(g)$ and hydrocarbons, two different catalysts are required for optimum performance. For example, $CuO(s)$ and $Cr_2O_3(s)$ may be used to catalyze the combustion of $CO(g)$ and hydrocarbons. These catalysts first adsorb the oxygen gas in the exhaust; this weakens the O—O bond in $O_2(g)$ and provides oxygen atoms to react with the adsorbed $CO(g)$ to form $CO_2(g)$. Oxidation of the hydrocarbons probably occurs in a similar manner, with the hydrocarbons first being adsorbed and then by weakening a C—H bond. There are problems with the catalytic converter. For example, the exhaust from automobiles contains traces of $SO_2(g)$, and the conversion of $SO_2(g)$ to $SO_3(g)$ is catalyzed by the converter. The $SO_3(g)$ may then dissolve in water to form sulfuric acid, which is a dangerous pollutant.

3.7 Half-life, residence time, and renewal time

The *half-life* ($t_{1/2}$) of a reactant is defined as the time required for the concentration of the reactant to decrease to halfway between its initial and final values. We can derive an expression for $t_{1/2}$ for a reactant A for a reaction that is first order in A by substituting $[A]_t = \frac{1}{2}[A]_0$ and $t = t_{1/2}$ in Eq. (3.5), which yields

$$\log\frac{\frac{1}{2}[A]_0}{[A]_0} = -\frac{kt_{1/2}}{2.303}$$

or,

$$t_{1/2} = \frac{2.302\log 2}{k}$$

therefore,

$$t_{1/2} = \frac{0.693}{k} \tag{3.10}$$

Equation (3.10) shows that for a first-order reaction, $t_{1/2}$ is independent of the initial concentration of the reactant.

Chemicals are constantly being discharged into the oceans and into the atmosphere; they are also produced by chemical reactions in seawater and air, yet the overall chemical compositions of the oceans and the atmosphere do not change greatly (although there are some important exceptions). This is because there are *sinks* that remove trace chemicals at about the same rate as they are injected or produced, so that most chemicals exist in roughly steady-state conditions in large reservoirs such as the ocean or atmosphere. An important parameter related to a chemical under steady-state conditions is its *residence time* (τ) in the system, which is

$$\tau = \frac{M}{F} \tag{3.11}$$

where M is the amount (say in m^3) of the chemical in the reservoir and F the influx (i.e., rate of input plus rate of production) of the chemical to the reservoir (in $m^3\ s^{-1}$). If M and F change with time

$$\tau_t = \frac{M_t}{F_t} \tag{3.12}$$

where the subscript t indicates the value at time t. We could also define, in an analogous way, the residence time in terms of the efflux (i.e., rate of removal plus rate of destruction) of a chemical from a reservoir.

A useful analogy here is a tank of water, which can represent the reservoir. Suppose the tank is full of water and overflowing at its top due to water being pumped into the bottom of the tank at a rate F. If we assume that the water entering the bottom of the tank steadily displaces the water lying above it, by pushing it upward without any mixing, the time that each small element of water that enters the bottom of the tanks spends in the tank, before overflowing at the top, is M/F, where M is the volume of the tank [hence the reason for defining residence time as M/F in Eq. (3.11)]. In this case, when no mixing occurs in the reservoir, the residence time of the water is the same as the *renewal time* (T), which is defined as the time required to completely displace the original water from the tank.

Now let us consider a more realistic situation for natural systems; namely, one in which mixing takes place between the material that is injected into the reservoir and the material that is already residing in the reservoir. For simplicity, we will consider the mixing to be complete and thorough (i.e., perfect mixing). The tank representation of our reservoir is again helpful. Suppose that at time zero the tank is full of dirty water, and at this time clean water starts to be pumped into the bottom of the tank. Since the mixing is perfect, the rate of removal of dirty water from the top of the tank will be proportional to the fraction of the water in that tank that is dirty water. Therefore, if W is the amount of dirty water in the tank at time t

$$\frac{dW}{dt} = -kW \tag{3.13}$$

where k is a constant of proportionality. From Eqs. (3.11) and (3.13) we have for the dirty water

$$\tau = \frac{W}{(-dW/dt)} = \frac{1}{k} \tag{3.14}$$

Since Eq. (3.13) has the same form as Eq. (3.4), the half-life of the dirty water is given by Eq. (3.10). Combining Eqs. (3.10) and (3.14), we obtain the following relationship between the half-life $(t_{1/2})$ and the residence time (τ) for the case of perfect mixing

$$t_{1/2} = 0.693\tau \tag{3.15}$$

In the case of perfect mixing, the renewal time (T) will be infinitely long, since some molecules of dirty water will always be present in the tank. However, we can obtain an approximate relationship between T, $t_{1/2}$ and τ for perfect mixing as follows. From the definition of the half-life $(t_{1/2})$, we know that after $t_{1/2}$ minutes one-half of the dirty water will be left in the tank, and after $2t_{1/2}$ minutes $(1/2)(1/2) = (1/2)^2$ of the dirty water will be left in the tank, etc. Therefore, after 6 $t_{1/2}$ minutes $(1/2)^6 = (1/64)$ of the dirty water will be left in the tank. If we (arbitrarily) decide that 1/64 is a sufficiently small fraction that most of the dirty water can be considered to have been displaced, for a chemical that is perfectly mixed in a reservoir for which the efflux is given by a first-order reaction [Eq. (3.13)] we have the following relationships between the renewal time (T), the half-life ($t_{1/2}$) and the residence time (τ)

$$T \approx 6t_{1/2} \approx 4\tau \tag{3.16}$$

In practice, of course, natural systems will fall somewhere between the cases considered above of no mixing and perfect mixing.

In the ocean, elements that form insoluble hydroxides have relatively short residence times (e.g., Al and Fe have residence times in the ocean of 100 and 200 years, respectively). Cations, such as $Na^+(aq)$ and $K^+(aq)$, and anions, such as $Cl^-(aq)$ and $Br^-(aq)$, have longer residence times in the ocean ($\sim 7 \times 10^6$ to 10^8 years). In the atmosphere, the very stable gas nitrogen has a residence time of a million years or so, while oxygen has a residence time of 5,000–10,000 years. Sulfur dioxide, water, and carbon dioxide, on the other hand, have residence times in the atmosphere of only a few days, 10 days, and 4 years, respectively. Of course, residence times may be determined by physical removal processes (e.g., scavenging by precipitation) as well as chemical.

Exercises

3./. Answer, interpret, or explain the following in light of the principles discussed in this chapter.

(a) A piece of paper can remain in contact with air indefinitely without any observable reaction, but it reacts as soon as it is touched with a flame.

(b) A match is stable until its head is rubbed on a rough surface.

(c) Increasing pressure generally decreases the rates of reactions of solids or liquids.

(d) Increasing pressure always increases the rates of reactions of gases.

(e) A catalyst is often in the form of a fine powder.

(f) A catalyst for a forward reaction has an equal effect on the reverse reaction.

(g) Gases that have small residence times in the atmosphere tend to have large spatial variabilities.

3.8.   The initial reaction rate of a hypothetical reaction

$$A + B \rightarrow G$$

was measured at 25°C for several different starting concentrations of A and B. The results are shown below.

Exper- iment number	[A] (M)	[B] (M)	Initial reaction rate (M s^{-1})
1	0.20	0.30	2.0×10^{-6}
2	0.40	0.30	1.6×10^{-5}
3	0.20	0.60	4.0×10^{-6}

Use these data to determine the values of m, n and k in Eq. (3.3) for this reaction. What will be the reaction rate at 25°C for [A] = 0.10 M and [B] = 0.80 M?

3.9.   Show that if the overall order of a chemical reaction is zero, the concentration $[A]_t$ of a reactant at time t is given by

$$[A]_t = -akt + [A]_0$$

where a is the coefficient of A in the equation for the balanced chemical reaction, k the rate coefficient, and $[A]_0$ the initial concentration of A.

3.10.   The second-order rate coefficient for the reaction

$$NO(g) + O_3(g) \rightarrow NO_2(g) + O_2(g)$$

is 1.8×10^{-20} m^3 s^{-1} molecule^{-1} at 0°C. (a) Write an expression for the pseudo first-order rate coefficient for the above reaction in air in terms of $[O_3(g)]$. (b) If the concentration of ozone in air is constant at 15 parts per billion by volume (ppbv), determine the value of this pseudo rate coefficient at 1 atm and 0°C.

3.11.   For a first-order reaction involving gases that is of the form

$$A(g) \rightarrow products,$$

show that

$$\log(p_A)_t = \frac{-k}{2.303}t + \log(p_A)_o$$

where $(p_A)_o$ and $(p_A)_t$ are the partial pressures of A at the start of the reaction and at time t, respectively, and k is the rate coefficient for the reaction.

3.12.   The following mechanism has been proposed for the conversion of $O_3(g)$ into $O_2(g)$

$$O_3(g) \rightleftarrows O_2(g) + O(g) \qquad \text{(i)}$$
$$O_3(g) + O(g) \rightarrow 2O_2(g) \qquad \text{(ii)}$$

(a) What is the overall chemical reaction?
(b) What is the intermediate?
(c) What is the rate law for each step?
(d) If the experimentally determined rate law for the overall reaction is

$$\text{Rate} = k[O_3(g)]^2[O_2(g)]^{-1}$$

what is the rate-controlling step?
(e) On what would you surmise that $[O(g)]$ depends?

3.13.   The experimentally determined rate law for the reaction

$$2NO(g) + O_2(g) \rightarrow 2NO_2(g)$$

is

$$-\frac{d[O_2(g)]}{dt} = k[NO(g)]^2[O_2(g)]$$

Show that the reaction mechanism

$$NO(g) + O_2(g) \rightleftarrows OONO(g) \text{ (fast equilibrium)}$$
$$NO(g) + OONO(g) \rightarrow 2NO_2(g) \text{ (slow)}$$

is consistent with the rate law. Show that the alternative steps

$$NO(g) + NO(g) \rightleftarrows N_2O_2(g) \text{ (fast equilibrium)}$$
$$N_2O_2(g) + O_2(g) \rightleftarrows 2NO_2(g) \text{ (slow)}$$

are also consistent with the rate law. [Additional measurements show that the concentration of $OONO(g)$ is more

important than $N_2O_2(g)$ during the reaction, which establishes that the first mechanism is more important than the
second.]

3.14.   The elementary reaction

$$NO_2(g) + NO_3(g) \rightleftharpoons N_2O_5(g)$$

which is a possible source of cloud-water nitrate, occurs in
a single step. If the rate coefficients for the forward and
reverse reactions are 1.5×10^{-12} and 5.0×10^{-2} M^{-1} s^{-1},
respectively, what is the equilibrium constant for the re-
action?

3.15.   The overall reaction for nitramide (O_2NNH_2) (a possible
constituent of secondary particles formed by urban air pollution) decomposing in solution is

$$O_2NNH_2(aq) \rightarrow N_2O(aq) + H_2O(aq)$$

The rate law determined experimentally is

$$\frac{d[N_2O(aq)]}{dt} = k\frac{[O_2NNH_2(aq)]}{[H^+(aq)]}$$

Some proposed reaction mechanisms are

$$O_2NNH_2(aq) \underset{k_{1r}}{\overset{k_{1f}}{\rightleftharpoons}} N_2O(aq) + H_2O(aq) \qquad \text{(slow)} \qquad \text{(i)}$$

$$O_2NNH_2(aq) + H^+(aq) \underset{k_{2r}}{\overset{k_{2f}}{\rightleftharpoons}} O_2NNH_3^+(aq) \quad \text{(fast equilibrium)}$$

$$\qquad\qquad\qquad\qquad\qquad\qquad\qquad\qquad\qquad\qquad\qquad \text{(ii)}$$

$$O_2NNH_3^+(aq) \underset{k_{3r}}{\overset{k_{3f}}{\rightleftharpoons}} N_2O(aq) + H_3O^+(aq) \qquad \text{(slow)}$$

$$O_2NNH_2(aq) \underset{k_{4r}}{\overset{k_{4f}}{\rightleftharpoons}} O_2NNH^-(aq) + H^+(aq) \quad \text{(fast equilibrium)}$$

$$O_2NNH^-(aq) \underset{k_{5r}}{\overset{k_{5f}}{\rightleftharpoons}} N_2O(aq) + OH^-(aq) \qquad \text{(slow)} \qquad \text{(iii)}$$

$$H^+(aq) + OH^-(aq) \underset{k_{6r}}{\overset{k_{6f}}{\rightleftharpoons}} H_2O(aq) \qquad \text{(fast)}$$

(a) Which of the reaction mechanisms is consistent with the
rate law?

(b) What is the relationship between k and the rate coefficients for the reaction mechanism chosen in question (a)?

(c) What is the relationship between the equilibrium constant (K_c) for the overall reaction and the rate coefficients for the reaction mechanism chosen in question (a)?

3.16.　Calculate the approximate *maximum* rate for a gaseous bimolecular chemical reaction at 0.50 atm and $-20°C$, given that the molecular diameter (s) is 3.0×10^{-10} m and the average speed of a molecule ($\bar{c}$) is 481 m s^{-1}? How long would it take to consume the reacting gases at this rate? (Compare the answers with the solutions to Exercise 3.5.)

3.17.　By inspecting and generalizing Figure 3.2, show that

$$\Delta H_{rx} = E_a \text{ (forward)} - E_a \text{ (reverse)}$$

Hence, show that for an endothermic reaction, the activation energy in the forward direction will, in general, exceed the heat of reaction.

3.18.　Many reaction rates roughly double when the temperature increases by about 10°C above room temperature (20°C). What is the activation energy of a reaction that obeys this rule exactly?

3.19.　The activation energy for a reaction is 65 kJ mol^{-1}. What is the effect on the rate of the reaction of changing the temperature from 25°C to 32°C?

3.20.　The following table gives values of the rate coefficient (k) at various temperatures (T) for the reaction

$$N_2O_5(g) \rightarrow 2NO_2(g) + \tfrac{1}{2} O_2(g)$$

By plotting log k against $1/T$ determine the activation energy (E_a) and the frequency factor (A) for this reaction.

T(K)	k(s^{-1})	T(K)	k(s^{-1})
338	4.87×10^7	308	1.35×10^6
328	1.50×10^7	298	3.46×10^5
318	4.98×10^6	273	7.87×10^3

3.21.　The half-life of a first-order reaction is 20.2 s. How much of a 10.0-g sample of an active reactant in the reaction will be present after 60.6 s?

3.22.   Show that for a reaction that is second-order in just one reactant A, the half-life of A is $1/k''[A]_0$. (*Note:* In this case, the half-life depends on the initial concentration of A for every half-life interval.)

3.23.   A wooden carving, found on an archaeological site, is subjected to radiocarbon dating.[7] The carbon-14 activity is 12 counts per minute per gram of carbon, compared to 15 counts per minute per gram of carbon for a living tree. What is the maximum age of the carving? (*Hint:* Radioactive decay, as measured by the activity in counts per minute per gram of sample, is directly proportional to the number of radioactive atoms present in the sample. Therefore, it is described by the relations for a first-order chemical reaction given in Sections 3.1 and 3.7. The half-life of carbon-14 is 5.7×10^3 yr.)

3.24.   Ammonia (NH_3), nitrous oxide (N_2O), and methane (CH_4) comprise 1.0×10^{-7}, 3.0×10^{-5}, and 1.7×10^{-4}% by mass of the Earth's atmosphere, respectively. If the effluxes of these chemicals from the atmosphere are 1.0×10^{11}, 1.0×10^{11}, and 7.5×10^{11} kg yr^{-1}, respectively, what are the residence times (with respect to their effluxes) of NH_3, N_2O, and CH_4 in the Earth's atmosphere? (Mass of the Earth's atmosphere $= 5.14 \times 10^{18}$ kg.)

3.25.   A lake has an area of 10 hectares and is 6 m deep. The lake initially contains 5% by volume of a liquid chemical X. If more of the chemical X is put into the lake at a rate of 500 m^3 per minute, what is the initial (instantaneous) residence time of chemical X in the lake with respect to (a) the influx of X and (b) the efflux of X? Assume that, due to seepage, the level of the lake does not change when liquid is put into it at a rate of 500 m^3 per minute.

Notes

1 "M" in the equation for a chemical reaction indicates any molecule that can take up excess energy that needs to be removed in order for the reaction to proceed. Thus, in the example given, the molecule M can collide with the OH and SO_2 molecules and reduce their energies so that they can combine to form $HOSO_2$.

2 In practice, the principle of detailed balancing works moderately well as an approximation; it is better for slow reactions than fast.

3 A strict mathematical treatment of this collision problem, in which the motion of all the molecules is taken into account, gives a collision rate per molecule of $\frac{4}{3}\pi\rho^2\bar{c}n$.

4 Stated more strictly, the activation energy is equal to the total energies that the reactants must possess in order for the reaction to occur. Thus, the rotational and vibrational energies of the molecules, as well as their translational energies, can contribute to the activation energy.

5 Chemists use the term *homogeneous* and *heterogeneous* differently than physicists. For example, in cloud physics, a homogeneous process is one involving just one substance (in any phase), and a heterogeneous process involves more than one substance.

6 Do not confuse *adsorption* with *absorption*. The former term refers to the binding of molecules to a surface, and the latter to the inclusion of molecules into the interior of another substance.

7 Radiocarbon dating of organic materials is based on the following principles. Carbon-12 [i.e. carbon with a mass number (= number of protons + number of neutrons) of 12] is the stable isotope of carbon. Carbon-14 is unstable (i.e., radioactive) with a half-life of 5,700 yr. Because carbon-14 is produced in the upper atmosphere, the ratio of carbon-14 to carbon-12 in the atmosphere is constant (and is believed to have been so for at least 50,000 yr). Carbon-14 is incorporated into atmospheric carbon dioxide, which is in turn incorporated, through photosynthesis, into plants. When animals eat plants, the carbon-14 is then incorporated into their tissues. While a plant or animal is alive, it has a constant intake of carbon compounds, and it maintains a ratio of carbon-14 to carbon-12 that is identical to that of the air. When a plant or animal dies, it no longer ingests carbon compounds. Therefore, the ratio of carbon-14 to carbon-12 decreases with time, due to the radioactive decay of carbon-14. Hence, the period that elapsed since a plant or animal or organic material was alive can be deduced by comparing the ratio of carbon-14 to carbon-12 in the material with the corresponding ratio for air.

4

Solution chemistry and aqueous equilibria

4.1 Definitions and types of solutions

A *solution* is a homogeneous mixture of substances. For example, when salt dissolves in water, a homogeneous mixture, or solution, forms. The component of a mixture that is present in the greatest quantity or that determines the state of matter (solid, liquid, or gas) of the solution is called the *solvent*. The other component(s) is (are) called the *solute(s)*. If water is the solvent, the solution is said to be *aqueous* (abbreviation: *aq*). If the quantity of solute is relatively large, the solution is said to be *concentrated*; if it is relatively small, the solution is *dilute*. Although we generally think of solutions as being liquids, they may also be gases or solids: Air is a gaseous solution; alloys (e.g., steel) are solid solutions.

As the amount of solid solute dissolved in a liquid solvent increases, the reverse process, namely, the *crystallization* or *precipitation* of the solute from the solvent, becomes increasingly important. When the solute dissolves and precipitates at the same rate, the amount of solute in the solution will remain constant. The solution is then said to be *saturated,* and the amount of solute present in a given quantity of the saturated solution is called the *solubility* of the solute in the solvent. If a solution is below its solubility limit, it is said to be *unsaturated;* if it is above the solubility limit, it is said to be *supersaturated*. Supersaturated solutions are unstable, and may rapidly return to the equilibrium (saturated) state through the crystallization of an appropriate quantity of solute.

4.2 Solution concentrations

The amount of solute present in a given quantity of solvent (or total solution) is called the *concentration* of the solution. The concentration of a solution may be given as a mass/total mass percentage. For example, a 3% NaCl solution by mass means 3 g of NaCl dissolved in 97 g of pure water. It may also be given as a volume/total volume percentage. Therefore, a 20% alcohol–water solution by volume is 20 mL of alcohol in 80 mL of pure water. Mass/volume percentage may also be used. If a solution concentration is given as simply a percentage, it may be assumed it is a mass/total mass percentage.

The preferred unit for the concentration of a solution in the SI system is *molar concentration* (or *molarity*) which is defined by

$$\text{Molar concentration} = \frac{\text{Number of moles of solute}}{\text{Number of liters of solution}} \quad (4.1)$$

The units of molarity are mol L^{-1} (or M for short). For example, a 2.5 M solution contains 2.5 gram moles of solute in every liter of solution.

Since the volume of a solution changes with temperature, so will the molarity, even though the amount of solute remains the same. If, however, the concentration of a given solution is expressed as moles of solute per kilogram of solvent (called *molal concentration* or *molality*), it will be independent of temperature. The units of molality are mol kg^{-1} (or m, for short).

The concentration of a particular solute i in a solution can also be expressed as a mole fraction

$$\psi_i = \frac{\text{Number of moles of i}}{\text{Total number of moles of all compounds in the solution}} \quad (4.2)$$

where ψ_i is dimensionless.

Exercise 4.1. Assuming that dry air consists of just molecular nitrogen, molecular oxygen, and atomic argon, and that these three gases contribute 75.5, 23.2, and 1.30 kg to 100 kg of air, calculate their mole fractions in air.

Solution. Since the molecular weights of nitrogen, oxygen, and argon are 28.0, 32.0, and 39.9, respectively, the numbers of moles of nitrogen, oxygen, and argon in 100 kg of dry air are

$$\frac{75.5 \times 10^3}{28.0} = 27000, \quad \frac{23.3 \times 10^3}{32.0} = 725 \text{ and } \frac{1.30 \times 10^3}{39.9} = 32.6, \text{ respectively.}$$

Therefore, the total number of moles in 100 kg of dry air is $2700 + 725 + 32.6 = 3460$. Hence, from Eq. (4.2)

$$\psi_{nitrogen} = \frac{2700}{3460} = 0.780 = 78.0\%$$

$$\psi_{oxygen} = \frac{725}{3460} = 0.210 = 21.0\%$$

$$\psi_{argon} = \frac{32.6}{3460} = 0.00942 = 0.942\%$$

Exercise 4.2. The *apparent molecular weight* of solution M_a is defined as

$$M_a = \frac{\text{Total mass of a solution}}{\text{Total number of moles in solution}}$$

What is the apparent molecular weight of dry air? Use the assumptions and values given in Exercise 4.1.

Solution. From the definition of M_a and the values calculated in Exercise 4.1, the apparent molecular weight of dry air is

$$M_a = \frac{100 \times 10^3}{3460} = 28.9 \text{ g mole}^{-1}$$

4.3 Factors affecting solubility

The tendency toward maximum randomness (entropy) causes a solid to dissolve. On the other hand, the precipitation of a solid from a solution lowers the energy of the system, which is also a favored condition. Equilibrium is reached when the concentration of the solutes in a solution is such that the driving forces of these two opposing tendencies (randomness and minimum energy) are the same. The energy factor is measured by the change in heat content when 1 mole of a solid dissolves; this is called the *heat of solution*. An increase in temperature always favors the more random state (e.g., the dissolving process for a solid). Consequently, the solubilities of solids increase with increasing temperature.

Since gases are more random than liquids, randomness (or entropy) *decreases* as a gas dissolves in a liquid. Therefore, unlike solids, the tendency toward maximum randomness favors the gas phase rather than the gas dissolving in a liquid. However, when a gas dissolves in a liquid, heat is released, and this favors the dissolving process. As in the case of a solid, the equilibrium concentration (or solubility) of a gas in a liquid involves a balance between randomness and energy, but in the case of a gas these two driving forces act in the opposite directions to those in which they act on a solid. Consequently, for a gas, increasing the temperature, which favors the more random state, will *decrease* the solubility.

The solubilities of solids and liquids are not affected much by pressure. By contrast, the solubility of any gas in a solvent increases with increasing pressure. This is because increasing pressure decreases the randomness of the gas phase, and hence decreases the difference in randomness between the gas phase and the aqueous solution. Carbonated drinks (soda pop) depend on this fact. Bottled at pressure in excess of 1 atm, considerable quantities of CO_2 can be dissolved in the liquid. When the pressure is reduced to atmospheric by taking the cap off the bottle, the CO_2 rapidly comes out of solution in the form of bubbles.

An approximate relationship between pressure and solubility is given by *Henry's law*

$$C_g = k_H p \tag{4.3}$$

where C_g is the solubility of the gas in the solution, p the pressure of the gas over the solution, and k_H is a temperature-dependent proportionality constant called the *Henry's law constant*. Tabulated solubilities of gases are generally based on a gas pressure of 1 atm above the liquid.

4.4 Colligative properties

Some properties of a solution depend on the concentration of the solution but not on the particular identity of the solute. These are called *colligative* properties. Three colligative properties are discussed below: vapor-pressure lowering, boiling-point elevation, and freezing-point depression.

The effect of the concentration of a solution on the vapor pressure of the solvent is given approximately by *Raoult's law*, which states: *the vapor pressure (p_A) of a solvent (A) above a solution is equal to the product of the vapor pressure of the pure solvent p_A^0 and the mole fraction of the solvent in the solution (ψ_A).* That is,

$$p_A = \psi_A p_A^0 \tag{4.4}$$

Similarly, if the solute (B) is volatile

$$p_B = \psi_B p_B^0 \tag{4.5}$$

A solution that obeys Raoult's law exactly is called an *ideal solution*. Solutions that are associated with either exothermic or endothermic reactions are *not* ideal solutions. Raoult's law is most accurate when used to describe components of a solution that are present in high concentrations. At low concentrations there are often significant departures from Raoult's law. At very low concentrations the vapor pressure of a solute is given by Henry's law.

A liquid boils when its saturated vapor pressure is the same as the atmospheric pressure. Since a nonvolatile solute will lower the vapor pressure of a solution, a higher temperature will be required to cause the solution to boil. The increase in the boiling point of a solution (ΔT_B) above that of the pure solvent is approximately proportional to the molality (m)

$$\Delta T_B = K_b m \tag{4.6}$$

where K_b, the *molal boiling-point-elevation constant*, depends on the solvent. For water $K_b = 0.52°C\ m^{-1}$. Note that the elevation of the boiling point is proportional to the number of solute particles present in a given amount of solution.

The freezing point of a substance is the temperature at which the saturated vapor pressures of the solid and liquid phases are the same. Since solutes are not normally soluble in the solid phase of the solvent, the vapor pressure of the solid is unaffected by the solute. On the other hand, if the solute is nonvolatile, the vapor pressure of the solution is reduced. Consequently, the temperature at which the solution and solid phase will have the same saturated vapor pressure (i.e., the freezing point) is reduced. The reduction in the freezing point of a solution (ΔT_f) is given approximately by

$$\Delta T_f = K_f m \tag{4.7}$$

where K_f is called the *molal freezing-point-depression constant*. For water K_f is 1.86°C molality^{-1}.

4.5 Aqueous solutions; electrolytes

The central role that water plays on Earth is due not only to its great abundance but also to its unique ability to dissolve (at least partially) many substances. Consequently, aqueous solutions are extremely common in the atmosphere, plants, and animals (e.g., the human body consists primarily of aqueous solutions).

Solids can dissolve in water in two ways: (1) with their molecules intact (e.g., when sugar dissolves in water individual sugar molecules pass from the solid to the liquid phase but the sugar molecule does not break up), and (2) by their molecules breaking up into positively and negatively charged *ions*. For example, common salt (NaCl) dissolves in the latter way, which can be represented by

$$NaCl(s) \rightleftarrows Na^+(aq) + Cl^-(aq) \qquad (4.8)$$

where the plus sign indicates that the sodium ion carries one unit of positive charge, and the negative sign indicates that the chloride ion carries one unit of negative charge. (If an ion carries two units of positive charge, it is indicated by the superscript $2+$, etc.)

Aqueous solutions containing charged ions are electrically conducting and are called *electrolytes*. Aqueous ions are individual species, the properties of which are independent of their source. For example, chloride ions from NaCl are just the same as chloride ions from hydrochloric acid (HCl) or any other electrolyte containing chlorine.

A compound will dissolve in water to form ions if the attractive forces between the water molecule and the ions are stronger than the attractive force between the ions. For example, NaCl dissolves in water because the attractive forces between the water molecule and the $Na^+(aq)$ and $Cl^-(aq)$ ions are stronger than the attractive force between $Na^+(aq)$ and $Cl^-(aq)$. In a sodium chloride solution, water molecules surround both the $Na^+(aq)$ and $Cl^-(aq)$ ions, with the negatively charged end of the water dipole (i.e., the oxygen atom) pointed toward the $Na^+(aq)$ ions and the positively charged end of the water dipole (i.e., the hydrogen atoms) pointed toward the $Cl^-(aq)$ ions. This type of arrangement between solution and solvent molecules is known as *solvation,* and when the solvent is water as *hydration.*

Table 4.1 shows the solubilities of various compounds in water. A substance is considered to be soluble if it dissolves to produce a solution with a concentration of at least one-tenth of a mole per liter (0.1 M) at room temperature.

4.6 Aqueous equilibria

In Chapter 1 we discussed the basic principles of chemical equilibrium. We will now apply these principles to ionic equilibria in aqueous solutions.

Table 4.1. *Solubilities in water of some common compounds comprised of the negative and positive ions indicated*[a]

Negative ions (anions)	Positive ions (cations)	Solubility of compound
All	Li^+, Na^+, K^+, Rb^+, Cs^+, Fr^+ (alkali ions)	Soluble
All	H^+	Soluble
All	NH_4^+ (ammonium)	Soluble
NO_3^- (nitrate)	All	Soluble
CH_3COO^- (acetate)	All	Soluble
Cl^- Br^- I^-	Ag^+, Pb^{2+}, H_2^{2-}, Ca^+ All other	Low solubility Soluble
SO_4^{2-} (sulfate)	Ba^{2+}, Sr^{2+}, Pb^{2+}, Ca^{2+} All others	Low solubility Soluble
S^{2-} (sulfide)	Alkali ions, H^+, NH_4^+ Be^{2+}, Mg^{2+}, Ca^{2+}, Sr^{2+}, Ba^{2+} All others	Soluble Low solubility
OH^- (hydroxide)	Alkali ions, H^+, NH_4^+, Sr^{2+}, Ba^{2+} All others	Soluble Low solubility
PO_4^{3-} (phosphate) CO_3^{2-} (carbonate) SO_3^{2-} (sulfite)	Alkali ions, H^+, NH_4^+ All others	Soluble Low solubility

[a] Adapted from *Chemistry: An Experimental Science*, G. Pimentel, ed., copyright © 1963 by W. H. Freeman and Company. Reprinted with permission.

As a starting point, let us consider the dissolution of cuprous chloride (CuCl) in water

$$CuCl(s) \rightleftarrows Cu^+(aq) + Cl^-(aq)$$

The equilibrium constant for this reaction is

$$K_{sp} = [Cu^+(aq)][Cl^-(aq)] = 3.2 \times 10^{-7} \qquad (4.9)$$

(Recall that in the definition of the equilibrium constant the concentrations of solids and liquids are equated to unity – see Section 1.2). If C

is the molarity of cuprous chloride in water, C moles of cuprous chloride will dissolve in 1 L of water to produce C moles of the chloride ion (Cl^-) and C moles of the cuprous ion (Cu^+). Hence, $K_{sp} = 3.2 \times 10^{-7} = C^2$ or $C = 0.00057$ M. Therefore, only 5.7×10^{-4} moles of cuprous chloride dissolve in 1 L of water. (*Note:* We could have predicted from Table 4.1 that the solubility of CuCl in water is low.)

Since K_{sp} in Eq. (4.9) is the product of ion concentrations, it is called the *ion product constant* or *solubility product* (hence the subscript "sp"). Use of the solubility product is generally (and most accurately) used for substances with small solubilities. The solubility products for a number of salts and ions are listed in Appendix IV(c).

Exercise 4.3. Calculate the solubility of barium hydroxide, $Ba(OH)_2(s)$, in water at 25°C given that its solubility product is 5×10^{-3} at 25°C.

Solution. The reaction for the dissociation of barium hydroxide in water is

$$Ba(OH)_2(s) \rightleftharpoons Ba^{2+}(aq) + 2(OH^-(aq)) \tag{4.10}$$

The solubility product is

$$[Ba^{2+}(aq)] \, [OH^-(aq)]^2 = K_{sp} = 5 \times 10^{-3} \tag{4.11}$$

From Reaction (4.10) we see that two moles of OH^-(aq) are produced for every one mole of Ba^{2+}(aq). Therefore,

$$[OH^-(aq)] = 2[Ba^{2+}(aq)] \tag{4.12}$$

From Eqs. (4.11) and (4.12), $[Ba^{2+}(aq)]\{2[Ba^{2+}(aq)]\}^2 = 5 \times 10^{-3}$, or $4\,[Ba^{2+}(aq)]^3 = 5 \times 10^{-3}$. Therefore, $[Ba^{2+}(aq)] = 0.1$ M. Since, from Reaction (4.10), 1 mole of barium hydroxide dissolves for every 1 mole of Ba^{2+}(aq) that forms, the solubility of barium hydroxide in water is 0.1 M.

Even very small solubility products can be measured electrically, and these values are listed in chemical tables. As the above exercise illustrates, the solubility of a substance can be derived from its solubility product. Solubility products are generally listed only for slightly or sparingly soluble substances. If K_{sp} is very small, the substance is often termed *insoluble* (in water). In the case of moderately and highly soluble substances (such as NaCl or NaOH), the use of solubility products is not very useful. This is because instead of defining an equilibrium constant in terms of concentrations we would have to define it in terms of the *activities* of the substances (see note 5 in Chapter 1).

Whenever the product of the concentrations of any two ions (raised to the appropriate powers) in a solution is less than the corresponding K_{sp} value, the solution is *subsaturated*.

In Exercise 4.3 the ions in the saturated solution derive only from the solid solute. However, as illustrated in the next two problems, solubility products can be used to calculate the solubility of a substance in a saturated solution even if some of the ions derive from another source.

Exercise 4.4. (a) Calculate the molar solubility of silver chromate $[Ag_2CrO_4(s)]$ in water, given that the solubility product is 2.4×10^{-12}. (b) Discuss qualitatively the effects of adding some $CrO_4^{2-}(aq)$ ions from another source to a saturated solution of silver chromate in water.

Solution. (a) The reaction is

$$Ag_2CrO_4(s) \rightleftarrows 2Ag^+(aq) + CrO_4^{2-}(aq) \qquad (4.13)$$

The solubility product is

$$[Ag^+(aq)]^2[CrO_4^{2-}(aq)] = 2.4 \times 10^{-12} \qquad (4.14)$$

We see from Reaction (4.13) that when 1 mole of $Ag_2CrO_4(s)$ dissolves in water, 2 moles of $Ag^+(aq)$ and 1 mole of $CrO_4^{2-}(aq)$ appear in the saturated solution. Therefore,

$$[Ag^+(aq)] = 2[CrO_4^{2-}(aq)] \qquad (4.15)$$

From Eqs. (4.14) and (4.15)

$$\{2[CrO_4^{2-}(aq)]\}^2[CrO(aq)] = 2.4 \times 10^{-12}$$

Therefore, $[Cr_4^{2-}(aq)] = 0.84 \times 10^{-4}$ M. Since, from Reaction (4.13), 1 mole of silver chromate dissolves for every 1 mole of $CrO_4^{2-}(aq)$ that forms, the solubility of silver chromate in water is 0.84×10^{-4} M.

(b) If $CrO_4^{2-}(aq)$ ions are added to the equilibrium situation represented by Reaction (4.13), LeChatelier's principle tells us that the reverse reaction in (4.13) will be favored. This means that some $Ag_2CrO_4(s)$ will precipitate out from the saturated solution, thereby reducing the concentration of $Ag^+(aq)$.

We see from the last problem that the addition to a saturated solution of a small quantity of another solution that contains one of the ions of the saturated solution (called a *common ion*) reduces the solubility of the solute and causes precipitation of the excess solid.

Exercise 4.5. What is the molar solubility of $Ag_2CrO_4(s)$ in a 0.05 M solution of $K_2CrO_4(aq)$? The solubility product for $Ag_2CrO_4(s)$ is 2.4×10^{-12}. (Assume that $K_2CrO_4(aq)$ remains subsaturated as the $Ag_2CrO_4(s)$ is added.)

Solution. $K_2CrO_4(aq)$ acts as a source of $CrO_4^{2-}(aq)$ ions in the solution. Therefore, this is the common ion problem discussed qualitatively in Exercise 4.4. Let C be the molar solubility of $Ag_2CrO_4(s)$. Then, from Reaction (4.13), we have from the dissolving of the $Ag_2CrO_4(s)$, $[Ag^+(aq)] = 2C$ and $[CrO_4^{2-}(aq)] = C$. But there is a further 0.05 M of $CrO_4^{2-}(aq)$ from the $Ag_2CrO_4(s)$. We can summarize the situation as follows

$$Ag_2CrO_4(s) \rightleftharpoons \underbrace{2Ag^+(aq)} + \underbrace{CrO_4^{2-}(aq)}$$

From $Ag_2CrO_4(s)$:	$2C$	C	mol L^{-1}
From 0.05 M. of $K_2Cr_4(aq)$:	—	0.05	mol L^{-1}
Total concentrations:	$2C$	$C + 0.05$	mol L^{-1}

The usual K_{sp} relation must be satisfied. Therefore,

$$[Ag^+(aq)]^2[CrO_4^{2-}(aq)] + (2C)^2(C + 0.05) = 2.4 \times 10^{-12} \quad (4.16)$$

An exact solution for C would involve solving a cubic equation. However, we can make a simplification. We know from Exercise 4.4(b) that C will be less than it was in Exercise 4.4(a), that is, less than 0.84×10^{-4} M. Hence, $C + 0.05 \simeq 0.05$. Therefore, Eq. (4.16) becomes

$$4C^2(0.05) = 2.4 \times 10^{-12}$$

or,

$$C = 4 \times 10^{-6} \text{ M}$$

Hence, the equilibrium molar solubility of $Ag_2CrO_4(s)$ in a 0.05 M solution of $K_2CrO_4(aq)$ is 4×10^{-6} M. This is 21 times less than the solubility of $Ag_2CrO_4(s)$ in pure water which was calculated in Exercise 4.4(a).

Certain solutes produce a greater effect on colligative properties than expected from the relations given in Section 4.4. This can be allowed for empirically by introducing the *van't Hoff factor* (i), which is defined as

$$i = \frac{\text{Measured value of a colligative property}}{\text{Expected value of a colligative property}} \quad (4.17)$$

The following exercise illustrates how van't Hoff factors can be calculated and the reason for deviations from the expected values.

Exercise 4.6. A 1 M solution of NaCl in water causes an elevation in the boiling point of 1.04°C. What is the van't Hoff factor for NaCl?

Solution. From Eq. (4.6) we see that a 1 M solution is expected to produce an elevation in the boiling point of water of 0.52°C. Therefore, from Eq. (4.17) we have for NaCl

$$i = \frac{1.04}{0.52} = 2.0$$

The value of $i = 2.0$ for NaCl suggests that its enhanced effect on elevating the boiling point of water is due to its dissociation in water into two ions: $Na^+(aq)$ and $Cl^-(aq)$.

Exercise 4.7. A 1 M aqueous solution of acetic acid, $CH_3COOH(l)$, causes an elevation in the boiling point of 0.540°C. What is the van't Hoff factor for $CH_3COOH(l)$?

Solution. Proceeding as in Exercise 4.6, and assuming that the value of K_b in Eq. (4.6) is 0.520°C m^{-1},

$$i = \frac{0.540}{0.520} = 1.04$$

Now, if an acetic acid molecule dissociates, it produces two ions: $H^+(aq)$ and $CH_3COO^-(aq)$. Since the i value is only 1.04, this suggests that only about 4% of the molecules of acetic acid dissociate in water. This leads us to the subject of strong and weak electrolytes.

4.7 Strong and weak electrolytes; ion-product constant for water

If each molecule of a substance that dissolves in water dissociates into ions (rather than the molecule remaining intact in the water), the substance is called a *strong electrolyte*. Sodium chloride is a strong electrolyte [see Reaction (4.8)]; so is hydrochloric acid, HCl(g), each molecule of which when it enters water breaks up into two ions

$$HCl(g) \rightarrow H^+(aq) + Cl^-(aq) \quad (4.18)$$

Only one arrow is shown in Reaction (4.18), and it goes from left to right; this indicates that the ionization of HCl goes essentially to completion, and that the reverse reaction is negligible.

Weak electrolytes ionize incompletely in water. For example, acetic acid dissolves only partially in water (to form vinegar)

$$CH_3COOH(l) \rightleftharpoons H^+(aq) + CH_3COO^-(aq) \qquad (4.19)$$

but only a small fraction of the acetic acid forms ions in this way, so the solution is only a weak electrical conductor. The ionization of $CH_3COOH(l)$ represented by Reaction (4.19) differs from that of $HCl(g)$ in Reaction (4.18) in a significant way; namely, it is a truly reversible reaction [indicated by the two-way arrows in Reaction (4.19)] and can therefore be represented by an equilibrium-constant expression.

Water itself is a very weak electrolyte

$$H_2O(l) \rightleftharpoons H^+(aq) + OH^-(aq) \qquad (4.20)$$

where $H^+(aq)$ and $OH^-(aq)$ are the hydrogen and hydroxide ions, respectively, in solution. The longer arrow from right to left in Reaction (4.20) can, if desired, be used to emphasize that there are many more H_2O molecules than there are $H^+(aq)$ and $OH^-(aq)$ ions. We could define an equilibrium constant for Reaction (4.20) by

$$K = \frac{[H^+(aq)][OH^-(aq)]}{[H_2O(l)]} \qquad (4.21)$$

However, the concentration of H_2O molecules in water is so large (55.5 M), and so few ions are formed that $[H_2O(l)]$ is virtually constant. Therefore, $[H_2O(l)]$ is usually combined with K and called the *ion-product constant for water* (K_w)

$$K_w = [H^+(aq)] \, [OH^-(aq)] \qquad (4.22)$$

where,

$$K_w = [H_2O(l)]K = 55.5K \qquad (4.23)$$

At 25°C, $K_w = 1.00 \times 10^{-14}$ (when concentrations are expressed in moles per liter).

Exercise 4.8. What are the concentrations of $H^+(aq)$ and $OH^-(aq)$ in pure water at 25°C?

Solution. In pure water

$$[H^+(aq)] = [OH^-(aq)]$$

Combining this with Eq. (4.22)

$$K_w = [H^+(aq)]^2$$

Therefore,

$$[H^+(aq)] = [OH^-(aq)] = \sqrt{K_w} \qquad (4.24)$$

and, at 25°C

$$[H^+(aq)] = [OH^-(aq)] = \sqrt{1.00 \times 10^{-14}} = 1.00 \times 10^{-7} \text{ M}$$

From Exercise 4.8 we see that the ratio of $H^+(aq)$ ions to $H_2O(l)$ molecules in pure water at 25°C is only

$$\frac{1.00 \times 10^{-7} \text{ M}}{55.5 \text{ M}} = 1.80 \times 10^{-9}$$

In other words, there are only a few ions to every billion neutral water molecules. This is why pure water is only a weak electrolyte (i.e., its electrical conductivity is very small).

Exercises

4.9.   Answer, interpret, or explain the following in the light of the principles presented in this chapter.

(a) Give examples of liquid solutions in which the solvent is liquid and the solute is a gas, and in which the solvent is liquid and the solute is solid.

(b) The heat released when oxygen dissolves in water is 13 kJ mol^{-1}, and when nitrous oxide (N_2O) dissolves in water 20 kJ mol^{-1}. Assuming the randomness factor discussed in Section 4.3 is the same in both cases, which gas do you think would have the higher solubility in water?

(c) Soda pop goes "flat" if it warms up.

(d) If a glass of cold water warms up, bubbles of air form on the inside of the glass.

(e) Fish need oxygen dissolved in water to survive. What is the effect of heating rivers by industry ("thermal pollution") on fish?

(f) Ethylene glycol is used as an antifreeze.

(g) Calcium chloride is used to melt ice on roads.

(h) Lemons freeze at a higher temperatures than oranges.

(i) Although the solubility product (K_{sp}) and the molar solubility of a slightly soluble substance are related, they are never equal.

(j) How would you induce the precipitation of a soluble substance from a solution? Give an example.

4.10. A solution of density 0.88 g mL^{-1} consists of 9.8 g of toluene (C_7H_8) in 452 g of benzene (C_6H_6). What is the concentration of toluene in the solution as a mass/total mass percentage, as a molarity, as a molality, and as a mole fraction?

4.11. Show that if a solute dissolves in a liquid with an endothermic heat of solution, the solubility of the solute in the liquid will increase with increasing temperature.

4.12. Eighty cubic centimeters of hydrogen sulfide gas (H_2S), measured at STP, can dissolve in 18.3 g of water. What will be the molal concentration of a saturated solution of H_2S(aq) in water at 8.0 atm pressure?

4.13. A constant humidity can be produced in a small enclosure by placing in the enclosure an aqueous solution of the appropriate concentration. How much glycerin ($C_3H_9O_3$) would have to be placed in 1.00 kg of water to achieve a water vapor pressure of 12.0 mm of mercury at 20°C. (Saturated vapor pressure of water at 20°C is 17.5 mm of mercury.)

4.14. What proportions (by volume) of water and ethylene glycol ($C_2H_6O_2$) must be mixed to form a solution with a freezing point of − 10°C? (K_f = 1.86°C molality^{-1} for water; density of ethylene glycol is 1.12 g cm^{-3}.)

4.15. Which of the following compounds are soluble in water (i.e., have solubilities ≥ 0.1 M) and which have low solubilities: CuCl, AgI, Ba(OH)$_2$, KNO$_3$, H$_2$SO$_4$, Mg(OH)$_2$, and HCl?

4.16. Ammonium nitrate (NH$_4$NO$_3$) dissolves in water with an endothermic heat of reaction of 26.4 kJ mol^{-1}. How much ambient heat will be absorbed when 2.0 kg of ammonium nitrate dissolves in water? (This reaction is utilized in instant ice packs for treating injuries. Solid ammonia nitrate is contained inside a thin-walled plastic bag, which is sealed inside a thicker bag that also contains some water. The inner bag can be broken by pressing the outer bag, thereby allowing the ammonium nitrate to form a solution with the water which gets quite cold.)

4.17.   Write the expression for the solubility product for each of
the following reactions.

(a) $Ag_2CrO_4(s) \rightleftarrows 2Ag^+(aq) + CrO_4^{2-}(aq)$

(b) $CaSO_4(s) \rightleftarrows Ca^{2+}(aq) + SO_4^{2-}(aq)$

(c) $CH_3COOH(l) \rightleftarrows H^+(aq) + CH_3COO^-(aq)$

4.18.   The K_{sp} value for calcium carbonate ($CaCO_3$) is 2.8×10^{-9}
at 25°C. What is the solubility of $CaCO_3$ in water in grams
per liter?

4.19.   The concentration of $Ag^+(aq)$ in a solution is 2.0×10^{-4}
M. What concentration must $I^-(aq)$ reach before $AgI(s)$
precipitates? The solubility of $AgI(s)$ is 8.5×10^{-17}.

4.20.   Calculate the molar solubility of lead sulfate, $PbSO_4(s)$, in
(a) pure water, and (b) a 0.10 M solution of lead nitrate,
$Pb(NO_3)_2$. The solubility product for lead sulfate is
1.6×10^{-8}.

5

Acids and bases

Very early in the history of chemistry many substances were designated as *acids, bases,* and *salts.* Acids have a sour taste (e.g., citric acid gives lemon juice its sour taste); they dissolve certain metals; and they also dissolve carbonate minerals to produce carbon dioxide. Bases have a bitter taste (e.g., sodium carbonate); they feel slippery when touched; and they react with many dissolved metal salts to form precipitates. However, the most striking characteristic of bases is their ability to neutralize the properties of acids; when a base reacts with an acid a salt is produced.

The French chemist Lavoisier thought that all acids contain oxygen (the word oxygen means "acid former" in Greek). However, it was subsequently found that many acids contain no oxygen (e.g., hydrochloric acid, HCl), but that they all contain hydrogen.

Acids and bases figure prominently in the equilibrium of aqueous solutions, where they significantly enhance the electrical conductivity of water. In this chapter, we will explore some of the important properties of acids and bases in aqueous solutions. This will lead us to a discussion of several theories of acids and bases.

5.1 Some definitions and concepts

Equation (4.22) is valid for aqueous solutions as well as for pure water. A solution for which $[H^+(aq)] = [OH^-(aq)]$ is said to be *neutral;* thus, pure water is neutral. If $[H^+(aq)] > [OH^-(aq)]$ the solution is said to be *acidic.* If $[OH^-(aq)] > [H^+(aq)]$ the solution is said to be *basic.*

Exercise 5.1. If 0.02 mole of hydrochloric acid is dissolved in 1 L of water, what are the concentrations of $H^+(aq)$ and $OH^-(aq)$ ions in the solution at 25°C?

83

Solution. Hydrochloric acid is a gas under normal conditions, but it is a strong electrolyte that dissolves in water to form equal numbers of $H^+(aq)$ and $Cl^-(aq)$. Therefore, when 0.02 mole of HCl dissolve in 1 L of water, 0.02 mole of $H^+(aq)$ and 0.02 mole of $Cl^-(aq)$ will form [we can neglect the small additional number of $H^+(aq)$ produced by the dissociation of the water]. Hence, $[H^+(aq)] = 0.02$ M. Since we can apply Eq. (4.22) to the solution

$$[OH^-(aq)] = \frac{K_w}{[H^+(aq)]} = \frac{1.00 \times 10^{-14}}{0.02} = 5 \times 10^{-13} \text{ M}$$

Comparing the concentration of $OH^-(aq)$ calculated in the above exercise with that calculated for pure water in Exercise 4.8, we see that by adding 0.02 mole of HCl to 1 L of water, we have lowered the $OH^-(aq)$ concentration from 10^{-7} M to 5×10^{-13} M! But the $H^+(aq)$ concentration has increased from 10^{-7} M to 0.02 M. Clearly, the solution is now very acidic. Thus, HCl is both a strong electrolyte and a strong acid, because when a small amount of it is added to water it produces a strongly acidic solution. Similarly, a substance that dissolves in water to produce relatively high concentrations of $OH^-(aq)$ ions (compared to those in pure water), such as NaOH(s), will raise the concentration of $OH^-(aq)$ relative to $H^+(aq)$ by a tremendous amount (see Exercise 5.8). Such materials are called *bases,* because when added to water they form basic solutions.[1] It follows from the above definitions that acids and bases have opposite (or opposing) tendencies. Thus, bases react with acids to neutralize their acidity. For example, two neutralizing reactions are

$$HCl + NaOH \rightarrow H_2O + NaCl$$
$$H_2SO_4 + CaO \rightarrow H_2O + CaSO_4$$

In both cases, an acid reacts with a base to produce water and a third class of substance called a *salt* (e.g., NaCl and CaSO$_4$).

The acidity (or alkalinity, as the case may be) of water is very important because $H^+(aq)$ and $OH^-(aq)$ ions play crucial roles in many reactions in aqueous solutions. For example, the acidity (or alkalinity) determines the ability of water to sustain fish and plant life; it also determines the solubility of many materials in water.

In addition to the acids we have already mentioned, some other common acids are sulfuric acid (H_2SO_4), nitric acid (HNO_3), formic acid (HCOOH), phosphoric acid (H_3PO_4), hydrogen fluoride (HF), hydrobromic acid (HBr), hydroiodic acid (HI), and perchloric acid ($HClO_4$). Based on the discussion so far, we might hypothesize that acids have the following properties in common: they contain hydrogen, they dissolve in water to produce ions that conduct electricity (i.e., they are electrolytes), and one of the ions they release is H^+(aq).

Some common bases are sodium hydroxide (NaOH), potassium hydroxide (KOH), magnesium hydroxide ($Mg(OH)_2$), calcium oxide (CaO), sodium carbonate (Na_2CO_3), and ammonia (NH_3). We could postulate that, like acids, a common property of bases is their ability to dissolve in water to produce ions. Also, since bases counteract acids, we might conclude that one of the ions they produce can remove the hydrogen ion. For NaOH, KOH, and $Mg(OH)_2$ this ion is clearly OH^-(aq). But what is it for Na_2CO_3 and NH_3? To answer this question let us consider what happens when ammonia dissolves in water. It reacts with water molecules to form ammonium ions, NH_4^+(aq),

$$NH_3(aq) + H_2O(l) \rightleftharpoons NH_4^+(aq) + OH^-(aq) \qquad (5.1)$$

The ammonia also reacts with the H^+(aq) in water

$$NH_3(aq) + H^+(aq) \rightleftharpoons NH_4^+(aq) \qquad (5.2)$$

We see that Reaction (5.1) releases OH^-(aq), which can remove H^+(aq). For Na_2CO_3(s) the reactions in water are

$$Na_2CO_3(s) \rightleftharpoons 2Na^+(aq) + CO_3^{2-}(aq) \qquad (5.3)$$

followed by

$$CO_3^{2-}(aq) + H_2O \rightleftharpoons HCO_3^-(aq) + OH^-(aq) \qquad (5.4)$$

which also releases OH^-(aq). Therefore, we might describe a base as a substance that produces (either directly or indirectly) OH^-(aq) ions.

The above description of acids and bases, in which H^+(aq) and OH^-(aq) ions are viewed as responsible for acidic and basic properties, respectively, and different acidic (or electrolytic) strengths are attributed to varying degrees of ionic dissociation, was developed by the Swedish chemist S. Arrhenius between 1880 and 1890. While very useful, this theory has some problems. The first problem has to do with the nature of the positive-charge carrier in aqueous solutions; the second problem is that some substances can act as bases, even though they do not release OH^-(aq) ions. We will now consider both of these problems.

5.2 The nature of $H^+(aq)$

In Section 4.5 we pointed out that water is an excellent solvent for ionic compounds because of the strong attractive forces exerted by the water molecules. Thus, each ion in aqueous solution may be attached to several (four to six) water molecules. We refer to this as *hydration*. Since the H^+ ion (or *proton*) is very small,[2] it should be hydrated to a much greater extent than other ions. Consequently, "free" protons are unlikely in aqueous solutions. For example, the hydrated proton H_3O^+ (consisting of one proton and a water molecule – called the *hydronium ion*) is very stable, and is more likely to exist in aqueous solutions than is H^+. Four water molecules attached to a proton ($H_9O_4^+$ or $H^+ \cdot 4H_2O$) might be even more stable. Unfortunately, the preferred form of the positive ion in aqueous solutions is not known. Therefore, if we wish to emphasize its likely hydrated form, we will indicate it by $H_3O^+(aq)$; other times we will indicate it as $H^+(aq)$. The notation "aq" itself can serve to remind us that all ions in water are extensively hydrated, and that the exact form of the ion may not be known.

5.3 The Brønsted–Lowry theory; conjugate acid–base pairs

In light of the above comments on the nature of $H^+(aq)$, instead of representing the dissociation of HCl in water by Reaction (4.18), we might better represent it by

$$HCl(aq) + H_2O(l) \rightleftharpoons H_3O^+(aq) + Cl^-(aq) \tag{5.5}$$

From this viewpoint, we could consider an acid as a substance that can donate a proton to another molecule (rather than as a substance that releases free protons).

The Arrhenius theory views all bases as substances that produce $OH^-(aq)$ ions. However, acid–base type reactions can occur in *non-aqueous* solvents, in which $OH^-(aq)$ cannot be present because there are no oxygen atoms in the system. For example, HCl reacts with pure liquid ammonia

$$HCl(aq) + NH_3(l) \rightleftharpoons NH_4^+(aq) + Cl^-(aq) \tag{5.6}$$

Since NH_3 has eliminated the acid HCl, we could consider NH_3 as a base.

These problems with the Arrhenius theory led J. Brønsted and T.

Lowry to propose a more general view of acids and bases, in which *acids tend to donate protons and bases tend to accept protons*. From this viewpoint, in both Reactions (5.5) and (5.6) HCl acts as an acid, and H_2O in Reaction (5.5) and NH_3 in Reaction (5.6) act as bases. As indicated by the two-way arrows in Reaction (5.5), $H_3O^+(aq)$ may donate a proton and $Cl^-(aq)$ may accept a proton. In this case, $Cl^-(aq)$ is the base and $H_3O^+(aq)$ the acid. Therefore, we could write

$$HCl(aq) + H_2O(l) \rightleftharpoons H_3O^+(aq) + Cl^-(aq) \qquad (5.7)$$
$$\text{acid 1 + base 2} \rightleftharpoons \text{acid 2 + base 1}$$

where, $HCl(aq)$) and $Cl^-(aq)$, which differ only by a proton, are called the *conjugate acid–base pair* for the forward reaction (indicated by 1), and $H_3O^+(aq)$ and $H_2O(l)$ are the conjugate acid–base pair for the reverse reaction (indicated by 2). In Reaction (5.6), HCl and Cl^- are the conjugate acid–base pair for the forward reaction, and NH_4^+ and NH_3 are the conjugate acid–base pair for the reverse reaction.

5.4 The Lewis theory

An even more general theory of acids and bases was given by the American chemist G. N. Lewis in 1923. In this theory, an acid is an electron acceptor and a base is an electron donor.[3] This is a more general theory than the Brønsted–Lowry theory, because it allows the acid–base classification to be applied to reactions in which neither $H^+(aq)$ nor $OH^-(aq)$ play a role, or even to reactions in which there is no solvent. For example, the following are acid–base reactions in the Lewis theory

$$Ag^+(aq) + 2CN^-(aq) \rightarrow Ag(CN)_2^-(aq)$$
$$\text{acid} \qquad \text{base}$$

$$Zn(s) + Cu^{2+}(aq) \rightarrow Zn^{2+}(aq) + Cu(s)$$
$$\text{base} \quad \text{acid}$$

5.5 Strengths of acids and bases; acid-dissociation (or ionization) constant

The Brønsted–Lowry view of acids and bases suggests that the strengths of acids can be compared by measuring their relative tendencies to release a proton to a common base (taken to be water). Thus, if we represent an acid by HA and consider its reaction with water

$$HA(aq) + H_2O(l) \rightleftharpoons H_3O^+(aq) + A^-(aq) \qquad (5.8)$$

we can measure the strength of HA by the equilibrium constant for the forward reaction of (5.8)

$$K_a = \frac{[H_3O^+(aq)][A^-(aq)]}{[HA(aq)]} \qquad (5.9)$$

K_a is called the *acid-dissociation (or ionization) constant* for HA.

The dissociation constant for the conjugate base A^- of HA is given by the equilibrium constant for the forward reaction of

$$A^-(aq) + H_2O(l) \rightleftharpoons HA(aq) + OH^-(aq) \qquad (5.10)$$

which is

$$K_b = \frac{[HA(aq)][OH^-(aq)]}{[A^-(aq)]} \qquad (5.11)$$

From Eqs. (5.9) and (5.11)

$$K_aK_b = [H_3O^+(aq)][OH^-(aq)] \qquad (5.12)$$

From Eqs. (5.12) and (4.22), remembering that $[H_3O^+(aq)] = [H^+(aq)]$, we see that

$$K_aK_b = K_w = 1.00 \times 10^{-14} \text{ (at 25°C)} \qquad (5.13)$$

where, K_w is the ion-product constant for water. Thus, *the product of the acid-dissociation constant for an acid and the base-dissociation constant for its conjugate base is equal to the ion-product constant for water.* Some values of acid-dissociation constants are given in Appendix IV. Values of the corresponding conjugate base-dissociation constants can be derived from the acid-dissociation constant using Eq. (5.13).

The strongest acid that can exist in water is $H^+(aq)$, because any material that releases a proton more readily than water will completely ionize in water to form $H^+(aq)$; for example, HCl, H_2SO_4, and HNO_3 are 100% ionized in water. Similarly, $OH^-(aq)$ is the strongest base that can exist in water, because any base that is a stronger proton acceptor will remove protons from the water to become what is called completely protonated [e.g., $S^{2-}(aq)$, $O^{2-}(aq)$ and $H^-(aq)$ are 100% protonated in water].

5.6 The pH scale

For dilute solutions, the molar concentrations of hydrogen ions are small. For example, the hydrogen ion concentration of a 0.1 M solution of HCl in water is 0.1 M. As a shorthand notation, the hydrogen ion concentrations of dilute solutions are generally indicated by their *pH value*,[4] which is defined by

$$pH = -\log[H^+(aq)] = -\log[H_3O^+(aq)] \qquad (5.14)$$

where concentrations are measured in moles per liter (M). Thus, a 0.1 M solution of HCl has a pH of $-\log(10^{-1}) = 1$. We see from definition (5.14) that (1) the greater the hydrogen ion concentration (i.e., the more acidic the solution) the smaller is the pH value of the solution, and (2) a change in the hydrogen ion concentration by an order of magnitude (e.g., from 10^{-1} to 10^{-2} M) changes the pH value by unity.

At the beginning of this section we defined a solution as being neutral if $[H^+(aq)] = [OH^-(aq)]$. Pure water is neutral; therefore, from Eqs. (5.12) and (5.13)

$$[H_3O^+(aq)][OH^-(aq)] = 10^{-14}$$

or,

$$[H_3O^+(aq)]^2 = 10^{-14}$$

Therefore, for pure water

$$[H_3O^+(aq)] = [H^+(aq)] = 10^{-7} \text{ M}$$

Hence, the pH of pure water is $-\log(10^{-7}) = 7$. It follows that acidic solutions have pH < 7 and basic solutions have pH > 7.

Observed pH values in nature are generally between 4 and 9. Seawater normally has a pH between 8.1 and 8.3. Streams in wet climates generally have a pH between 5 and 6.5 and in dry climates between 7 and 8. Soil water in the presence of abundant decaying vegetation may have a pH of 4 or lower. The pH of rainwater can range from quite acidic (around 4.0) in industrial regions to about 5.6 in very clean regions. We will discuss the acidity of rainwater in some detail at the end of this chapter, but the following exercise illustrates why even clean rainwater does not have a pH of 7.

Exercise 5.2. The pH of natural rainwater is about 5.6. Assuming that all of this acidity is due to the absorption of CO_2 by the rain, determine how many moles of CO_2 would have to be absorbed in 1 L of rainwater.

Solution. Since the pH of rainwater is 5.6, the concentration of $H_3O^+(aq)$ in natural rainwater is given by

$$pH = 5.6 = -\log[H_3O^+(aq)]$$

Therefore,

$$[H_3O^+(aq)] = 0.25 \times 10^{-5} \text{ M}$$

The main source of $H_3O^+(aq)$ when CO_2 dissolves in water is

$$CO_2(g) + H_2O(l) \rightleftharpoons H_2CO_3(aq) \qquad (5.15a)$$

$$H_2CO_3(aq) + H_2O(l) \rightleftharpoons HCO_3^-(aq) + H_3O^+(aq) \qquad (5.15b)$$

We see from Reactions (5.15) that for every mole of CO_2 that is absorbed in water, one mole of $H_3O^+(aq)$ is produced. Therefore, to produce 0.25×10^{-5} M of $H_3O^+(aq)$, 0.25×10^{-5} moles of CO_2 would have to be absorbed in each liter of rainwater. Since this is about the solubility of CO_2 in water at atmospheric pressure, we see that the absorption of CO_2 in rainwater will cause it to have a pH of about 5.6.

5.7 Polyprotic acids

In the above exercise we assumed that Reaction (5.15b) is the only source of $H_3O^+(aq)$ when CO_2 dissolves in water. In fact, following Reaction (5.15b) there is another source of $H_3O^+(aq)$, namely

$$HCO_3^-(aq) + H_2O(l) \rightleftharpoons CO_3^{2-}(aq) + H_3O^+(aq) \qquad (5.16)$$

We see from Reactions (5.15b) and (5.16) that H_2CO_3 (carbonic acid) contributes two protons to water. Substances that contribute more than one proton to water are called *polyprotic acids*. Other polyprotic acids are $H_2C_2O_4$ (oxalic acid), H_3PO_4 (phosphoric acid), and H_2SO_3 (sulfurous acid).

The successive acid-dissociation constants of a polyprotic acid are often labeled K_{a1}, K_{a2}, etc. Provided $K_{a1} \gg K_{a2}$, the second reaction can be neglected as a source of protons compared to the first reaction. This is the case for carbonic acid and is the justification for ignoring Reaction (5.16) in Exercise 5.2. Note that intuitively we would expect the neutral molecule $H_2CO_3(aq)$ to give up a proton more readily than the negatively charged ion $HCO_3^-(aq)$. Of course, if we were interested in the concentration of CO_3^{2-}(aq) in the solution, we would have to consider Reaction (5.16), since this is the only source of CO_3^{2-}(aq).

Values for the successive acid-dissociation constants for several polyprotic acids are given in Appendix IV.

5.8 Hydrolysis

A reaction between liquid water and an ionic species in the aqueous solution is called a *hydrolysis reaction*. For example, Reaction (5.16) is a hydrolysis reaction. Another example is

$$C_2H_3O_2^-(aq) + H_2O(l) \rightleftharpoons HC_2H_3O_2(aq) + OH^-(aq) \qquad (5.17)$$

The equilibrium constant for such a reaction is called the *hydrolysis constant* and is given the symbol K_h. Thus, the hydrolysis constant for Reaction (5.16) is

$$K_h = \frac{[CO_3^{2-}(aq)][H_3O^+(aq)]}{[HCO_3^-(aq)]}$$

Some important aspects of hydrolysis are illustrated by the following example. When the salt sodium acetate ($NaC_2H_3O_2$) is added to water, the pH rises above 7. This means that the concentration of OH^-(aq) in the solution increases relative to that of H_3O^+(aq). The reactions involved indicate why this is so

$$NaC_2H_3O_2(s) + H_2O(l) \rightleftharpoons Na^+(aq) + HC_2H_3O_2(aq) + OH^-(aq) \quad (5.18)$$

$$HC_2H_3O_2(aq) + H_2O(l) \rightleftharpoons C_2H_3O_2^-(aq) + H_3O^+(aq) \qquad (5.19)$$

$$C_2H_3O_2^-(aq) + H_2O(l) \rightleftharpoons HC_2H_3O_2(aq) + OH^-(aq) \qquad (5.20)$$

which, on addition, yields the net reaction

$$NaC_2H_3O_2(s) + 3H_2O(l) \rightleftharpoons$$
$$Na^+(aq) + 2OH^-(aq) + H_3O^+(aq) + HC_2H_3O_2(aq) \qquad (5.21)$$

Thus, two OH^-(aq) are produced for every one H_3O^+(aq).

The hydrolysis constant for the reaction of sodium acetate in water is, from reaction (5.20)

$$K_h = \frac{[HC_2H_3O_2(aq)][OH^-(aq)]}{[C_2H_3O_2^-(aq)]} \qquad (5.22)$$

We could think of Reaction (5.20) as the net reaction resulting from the sum of the following two reactions

$$C_2H_3O_2^-(aq) + H_3O^+(aq) \rightleftharpoons H_2O(l) + HC_2H_3O_2(aq) \qquad (5.23)$$

$$2H_2O(l) \rightleftharpoons H_3O^+(aq) + OH^-(aq) \qquad (5.24)$$

Net $C_2H_3O_2^-(aq) + H_2O(l) \rightleftharpoons HC_2H_3O_2(aq) + OH^-(aq)$

which is Reaction (5.20). [We see from Reaction (5.19) that $HC_2H_3O_2(aq)$ is the conjugate acid of $C_2H_3O_2^-(aq)$, because the ion can be converted to the acid by attaching an $H^+(aq)$.] Similarly, from Reaction (5.20), the acetate ion, $C_2H_3O_2^-(aq)$, is the conjugate base of acetic acid, $HC_2H_3O_2(aq)$.)

The equilibrium constant for the forward reaction of (5.23) is $1/K_a$, where K_a is the acid-dissociation (or ionization) constant of acetic acid. Also, the equilibrium constant for the forward reaction of (5.24) is the ion–product constant for water (K_w). Since the equilibrium constant for a net reaction is equal to the product of the equilibrium constants for the individual reactions (see Exercise 1.14c), the equilibrium constant for Reaction (5.20), namely K_h, is given by

$$K_h = \frac{K_w}{K_a} \qquad (5.25)$$

Equation (5.25) is a general relation. It shows that salts associated with weak acids (low values of K_a) and strong bases [e.g., the acid associated with sodium acetate is acetic acid, which is weak, and the base associated with sodium acetate is $OH^-(aq)$, which is strong] hydrolyze to produce basic solutions with large K_h values. For such salts, it is the *anion* that hydrolyzes [e.g., $C_2H_3O_2^-(aq)$ in Reaction (5.20)] to produce the $OH^-(aq)$ ions.[5] Similarly, salts that form the conjugates of strong acids and weak bases (e.g., NH_4Cl) hydrolyze to produce acidic solutions with large K_h values ($= K_w/K_b$, where K_b is the base-dissociation constant). For such salts, it is the *cation* that hydrolyzes [e.g., $NH_4^+(aq)$ for NH_4Cl] to produce the $H_3O^+(aq)$ ions. Salts that form the conjugates of strong acids and strong bases (e.g., $NaCl$) do not undergo hydrolysis, so they produce solutions with pH $= 7$. If a salt forms the conjugates of a weak acid and a weak base, the solution will be neutral, acidic, or basic depending on the relative values of K_a and K_b.

Exercise 5.3. What is the pH of a 0.10 M solution of NH_4Cl if the base-dissociation constant for NH_3 is 1.8×10^{-5}? What fraction of the NH_4Cl is hydrolyzed?

Solution. As noted above, NH_4Cl dissolves to form the conjugates of a weak base (NH_3) and a strong acid (HCl). Therefore, we would expect it to hydrolyze to produce an acidic solution. The reactions are

$$NH_4Cl(s) + H_2O(l) \rightleftarrows Cl^-(aq) + NH_4^+(aq) + H_2O(l)$$

$$NH_4^+(aq) + H_2O(l) \rightleftarrows NH_3(aq) + H_3O^+(aq)$$

The hydrolysis constant for the last reaction is

$$K_h = \frac{[NH_3(aq)][H_3O^+(aq)]}{[NH_4^+(aq)]} = \frac{K_w}{K_b}$$

where $K_w = 10^{-14}$ and K_b is the base-dissociation constant for $NH_3(aq)$, which is given as 1.8×10^{-5}. Therefore, $K_h = 5.6 \times 10^{-10}$. We see from the reaction equations that equal numbers of $NH_3(aq)$ and $H_3O^+(aq)$ are formed. If $x = [NH_3(aq)] = [H_3O^+(aq)]$, then $[NH_4^+(aq)] = 0.10 - x$. Hence,

$$K_h = 5.6 \times 10^{-10} = \frac{x^2}{0.10 - x}$$

Solving this as a quadratic equation and taking the positive root, we get $x = 0.75 \times 10^{-5}$ M. (*Note:* If we had anticipated that x would be small, we could have obtained the value of x much more quickly by writing $0.10 - x = 0.10$ in the above expression for K_h. Generally, it is quicker to make such approximations at the outset and then check to determine if the value of x that is obtained is small enough to justify the approximation.) The pH of the solution is

$$pH = -\log[H_3O^+(aq)] = -\log_{10}x$$

$$= -\log(0.75 \times 10^{-5}) = -(-5.1)$$

$$= 5.1 \text{ (acidic, as predicted)}$$

The fraction of $NH_4Cl(s)$ hydrolyzed is

$$\frac{\text{Amount of } NH_3(aq)}{\text{Original amount of } NH_4Cl(s)} = \frac{x}{0.10} = \frac{0.75 \times 10^{-5}}{0.10}$$

$$= 7.5 \times 10^{-5} \text{ or } 0.0075\%$$

5.9 Buffers

If the pH of a solution is not greatly affected by the addition of small amounts of acids or bases, the solution is said to be *buffered*. This will be the case if the solution contains a relatively large amount of an acid–base conjugate pair provided that neither the acid nor the base is very strong [e.g., acetic acid, $HC_2H_3O_2$, and the acetate ion, $C_2H_3O_2^-(aq)$]. If a small amount of a strong acid is added to such a solution, most of the added $H_3O^+(aq)$ will combine with the weak base of the buffer (to form the conjugate acid of that weak base), so that the $H_3O^+(aq)$ concentration and pH of the solution will not change very much. Similarly, if a small amount of a strong base is added to the solution, most of the $OH^-(aq)$ will combine with the weak acid of the buffer. From the equilibrium equation for the ionization of acetic acid, the reverse reaction of (5.23), we have

$$K_a \text{ (acetic acid)} = \frac{[C_2H_3O_2^-(aq)][H_3O^+(aq)]}{[HC_2H_3O_2(aq)]}$$

Therefore, to achieve a specific pH, or $[H_3O^+(aq)]$, the ratio of the concentration of acetic acid to acetate ion must be

$$\frac{[HC_2H_3O_2(aq)]}{[C_2H_3O_2^-(aq)]} = \frac{[H_3O^+(aq)]}{K_a(\text{acetic acid})}$$

We can generalize the above ideas as follows. Let us represent the weak acid in a buffer by HA and the corresponding salt by MA. The equilibrium reaction for the acid dissociation is

$$HA(aq) \rightleftarrows H^+(aq) + A^-(aq)$$

and the acid-dissociation constant is given by

$$K_a = \frac{[H^+(aq)][A^-(aq)]}{[HA(aq)]} \tag{5.26}$$

Or, since $A^-(aq)$ is the base and $HA(aq)$ the acid in the buffer,

$$[H^+(aq)] = \frac{[\text{acid}]}{[\text{base}]} K_a \tag{5.27}$$

From Eq. (5.26)

$$\log K_a = \log[H^+(aq)] + \log\frac{[A^-(aq)]}{[HA(aq)]}$$

or,

$$-\log K_a = pH - \log\frac{[A^-(aq)]}{[HA(aq)]}$$

Following the same convention as for pH, $-\log K_a$ can be written as pK_a. Therefore,

$$pH = pK_a + \log \frac{[\text{base}]}{[\text{acid}]} \qquad (5.28)$$

For most buffers pH $\simeq$ pK_a, and [base] $\simeq$ [acid]. Of course, the *capacity* of a buffer depends on the *amounts* of acid and base in the solution. The *buffering capacity* is defined as the number of moles of H^+ (or OH^-) that would have to be added to a solution to change its pH by one unit.

Exercise 5.4. One liter (1 L) of a buffer solution contains 0.10 mole of acetic acid, $HC_2H_3O_2$, and 0.10 mole of sodium acetate, $NaC_2H_3O_2$. If the acid-dissociation constant for acetic acid is 1.74×10^{-5}, what is the pH of the solution? What is the acid buffering capacity of the solution?

Solution. Since the molar concentrations of the acid and the base are equal, we have from Eq. (5.28)

$$pH = pK_a = -\log(1.74 \times 10^{-5})$$

or,

$$pH = 4.76$$

To lower the pH by one unit, enough H^+ must be added to convert $C_2H_3O_2^-$ to $HC_2H_3O_2$ such that

$$pH = 3.76 = pK_a + \log\left(\frac{0.10 - x}{0.10 + x}\right)$$

Therefore,

$$3.76 = 4.76 + \log_{10}\left(\frac{0.10 - x}{0.10 + x}\right)$$

and,

$$x = 0.081$$

Thus, 0.081 mole of H^+ would have to be added to the solution to lower its pH by one unit. The buffering capacity of the solution is therefore 0.081 mole.

Exercise 5.5. What will be the change in the pH of the buffer solution in Exercise 5.4 if 0.020 mole of hydrochloric acid is added?

Compare this with the change in pH when 0.020 mole of hydrochloric acid is added to 1 L of pure water.

Solution. Before the addition of the hydrochloric acid, HCl, the concentrations of acetic acid, $HC_2H_3O_2$, and acetate, $C_2H_3O_2^-(aq)$, in the solution are each 0.10 M. Because HCl is a strong acid, it reacts completely with the acetate to form acetic acid. Therefore, after the HCl is added, the concentrations of acetic acid and acetate are $(0.10 + 0.020)$ M and $(0.10 - 0.020)$ M, that is, 0.12 M and 0.080 M, respectively. Therefore, from Eq. (5.28), the new pH of the solution will be

$$pH = pK_a + \log\frac{[base]}{[acid]}$$
$$= -\log(1.74 \times 10^{-5}) + \log\frac{0.08}{0.12}$$
$$= 4.76 - 0.18 = 4.58$$

Therefore, the change in the pH of the buffer solution is $(4.76 - 4.58)$ or a decrease of 0.18 pH units. [Note that the $Na^+(aq)$ released from the sodium acetate combines with the $Cl^-(aq)$ to form NaCl, but this is neither acidic nor basic.] We can assume that when 0.020 mole of HCl dissolves in 1 L of pure water, 0.02 mole of $H^+(aq)$ is formed, and this is not neutralized. Therefore, the concentration of the $H^+(aq)$ is 0.020 M and the pH of the solution is

$$pH = -\log[H^+(aq)] = -\log(0.020) = 1.7$$

Since the pH of pure water is 7, the lowering of the pH in this case is 5.3 pH units. Comparing this with the decrease of 0.18 pH units for the buffer solution, we see the remarkable ability of a buffer to stabilize the pH.

5.10 Complex ions

Metal ions can act as Lewis acids (i.e., electron acceptors) toward water molecules that serve as Lewis bases (i.e., electron donors) and toward other Lewis bases. This can have a profound effect on the solubility of a metal salt. For example, AgCl(s) dissolves in aqueous ammonia because of the Lewis acid–base interaction between $Ag^+(aq)$ and $NH_3(aq)$

$$AgCl(s) \rightleftharpoons Ag^+(aq) + Cl^-(aq) \qquad (5.29)$$

$$Ag^+(aq) + 2NH_3(aq) \rightleftharpoons Ag(NH_3)_2^+(aq) \qquad (5.30)$$

Net: $AgCl(s) + 2NH_3(aq) \rightleftharpoons Ag(NH_3)_2^+(aq) + Cl^-(aq)$

For the Lewis base $NH_3(aq)$ to increase the solubility of $AgCl(s)$ – that is, to drive reaction (5.29) to the right – it must interact more strongly with $Ag^+(aq)$ than does water.

A metal ion combined with a Lewis base, such as $Ag(NH_3^+)_2$ (aq), is called a *complex ion*. The equilibrium constant for the formation of a complex ion from a metal ion in aqueous solution is called the *formation constant* (K_f) of the complex ion. The higher the value of K_f the more stable is the complex ion. Values of the formation constants for some metal complex ions are given in Appendix IV.

5.11 Mass balance and charge balance relations

The ionization of any weak Brønsted–Lowry acid HA is given by Reaction (5.8), which, for convenience, we repeat here

$$HA(aq) + H_2O(l) \rightleftharpoons H_3O^+(aq) + A^-(aq) \qquad (5.31)$$

Also,

$$2H_2O(l) \rightleftharpoons H_3O^+(aq) + OH^-(aq) \qquad (5.32)$$

How can we calculate $[HA(aq)]$, $[A^-(aq)]$, $[H_3O^+(aq)]$ and $[OH^-(aq)]$ for a given initial concentration of HA in water? Since there are four unknowns, we need four relationships between the unknowns.

The equilibrium constants for Reactions (5.31) and (5.32) provide two of the relationships

$$K_a = \frac{[H_3O^+(aq)][A^-(aq)]}{[HA(aq)]} \qquad (5.33)$$

and,

$$K_w = [H_3O^+(aq)][OH^-(aq)] \qquad (5.34)$$

where the values of K_a and K_w are assumed to be known.

We can obtain a third relationship from the fact that the mass of HA must be conserved. Therefore, we can write

$$[HA]_{initial} \simeq [HA(aq)] + [A^-(aq)] \qquad (5.35)$$

where $[HA]_{initial}$ is the amount of acid in the water *before* any dissociation occurs, and $[HA(aq)]$ and $[A^-(aq)]$ are the equilibrium concentrations of $HA(aq)$ and $A^-(aq)$ in Reaction (5.31). Note that Eq. (5.35) is written as an approximate relation because it assumes that the mass of hydrogen atoms that have dissociated from the $A^-(aq)$ is small compared to the masses of the species in Eq. (5.35). Equation (5.35) is called the *mass balance relation*.

The fourth relationship between the concentrations of the species in Reactions (5.31) and (5.32) expresses the fact that the solution is electrically neutral; therefore, the total concentration of positive ions must equal the total concentration of negative ions

$$[H_3O^+(aq)] = [A^-(aq)] + [OH^-(aq)] \qquad (5.36)$$

This is called the *charge balance relation*. In practice, the solution of Eqs. (5.33) through (5.36) may involve some laborious algebra, but judicious approximations can often simplify the solution, as illustrated in the next section.

5.12 The pH of rainwater

Exercise 5.2 suggested that the absorption of CO_2 in rainwater to form a weak solution of carbonic acid would give natural rainwater a pH of about 5.6. We can now consider this exercise quantitatively by using some of the concepts just introduced.

As we have seen, carbonic acid is polyprotic

$$H_2CO_3(aq) + H_2O(l) \rightleftarrows HCO_3^-(aq) + H_3O^+(aq) \qquad (5.37)$$

$$HCO_3^-(aq) + H_2O(l) \rightleftarrows CO_3^{2-}(aq) + H_3O^+(aq) \qquad (5.38)$$

with successive acid-dissociation constants at 25°C of $K_{a1} = 4.2 \times 10^{-7}$ and $K_{a2} = 5.0 \times 10^{-11}$. Therefore,

$$\frac{[H_3O^+(aq)][HCO_3^-(aq)]}{[H_2CO_3(aq)]} = K_{a1} = 4.2 \times 10^{-7} \qquad (5.39)$$

and,

$$\frac{[H_3O^+(aq)][CO_3^{2-}(aq)]}{[HCO_3^-(aq)]} = K_{a2} = 5.0 \times 10^{-11} \qquad (5.40)$$

Let us now calculate the concentration of $H_3O^+(aq)$, $H_2CO_3(aq)$, $HCO_3^-(aq)$, $OH^-(aq)$, and $CO_3^{2-}(aq)$ when CO_2 from the atmosphere dissolves in otherwise pure rainwater, given that the solubility of CO_2 in water is 1.0×10^{-5} M at 25°C and 1 atm. Since we have five unknowns we need five equations to solve this problem, and so far we have only two equations, namely (5.39) and (5.40). The other three equations are provided by the ion-product constant for water

$$[H_3O^+(aq)] \, [OH^-(aq)] = K_w = 1.0 \times 10^{-14} \tag{5.41}$$

the material balance relation

$$[H_2CO_3]_{initial} = 1.0 \times 10^{-5}M = $$
$$[H_2CO_3(aq)] + [HCO_3^-(aq)] + [CO_3^{2-}(aq)] \tag{5.42}$$

and the charge balance relation

$$[H_3O^+(aq)] = [HCO_3^-(aq)] + 2[CO_3^{2-}(aq)] + [OH^-(aq)] \tag{5.43}$$

where the 1.0×10^{-5} M in Eq. (5.42) follows from the fact that for every mole of CO_2 that dissolves in water one mole of H_2CO_3 is formed [see Reaction (5.15a)], and the coefficient 2 in Eq. (5.43) allows for the two units of negative charge on each $CO_3^{2-}(aq)$ ion.

The solution of Eqs. (5.39) through (5.43) is simplified if we make some approximations. Since $K_{a1} \gg K_{a2}$, the contributions to $[H_3O^+(aq)]$ from Reaction (5.38) is negligible compared to that from Reaction (5.37). Also, since the only source of $CO_3^{2-}(aq)$ is from Reaction (5.38), $[CO_3^{2-}(aq)]$ will be small compared to $[H_2CO_3(aq)]$ and $[HCO_3^-(aq)]$. Finally, since $OH^-(aq)$ derives only from the dissociation of water, and an acid has been added to the water, we can assume that $[H_3O^+(aq)] \gg [OH^-(aq)]$. Hence, from Eq. (5.43), $[H_3O^+(aq)] \simeq [HCO_3^-(aq)]$, and Eqs. (5.39) and (5.42) become

$$\frac{[H_3O^+(aq)]^2}{[H_2CO_3(aq)]} \simeq 4.2 \times 10^{-7} \tag{5.39a}$$

and,

$$1.0 \times 10^{-5} \simeq [H_2CO_3(aq)] + [H_3O^+(aq)] \tag{5.42a}$$

We now have two equations for the two unknowns, which yields: $[H_3O^+(aq)] \simeq 1.8 \times 10^{-6}$ M and $[H_2CO_3^-(aq)] \simeq 8.1 \times 10^{-6}$ M. Therefore, $[HCO_3^-(aq)] \simeq [H_3O^+(aq)] \simeq 1.8 \times 10^{-6}$ M. Substituting $[H_3O^+(aq)] \simeq 1.8 \times 10^{-6}$ M in Eq. (5.41) yields $[OH^-(aq)] \simeq 5.6 \times 10^{-9}$ M.

Finally, substituting values into Eq. (5.40) gives $[CO_3^{2-}(aq)] \approx 5.0 \times 10^{-11}$ M.[6]

We can now derive the pH of the above system from

$$pH = -\log[H_3O^+(aq)] \approx -\log(1.8 \times 10^{-6}) = 5.7$$

Therefore, rainwater (or any other water) that is exposed only to atmospheric CO_2 at 25°C and 1 atm will have a pH of about 5.7.

Of course, neither rainwater nor any other water is generally exposed to CO_2 alone. Even in regions of the globe well removed from sources of anthropogenic pollution, rainwater is exposed to natural SO_2 gas and sulfate particles in the air, which can decrease the pH of the rain to below 5.7. In the absence of bases, such as NH_3 (the main atmosphere gaseous base) and $CaCO_3$ (from soil dusts), which are often in low concentrations in natural air, the pH of rainwater can vary from about 4.5 to 5.6 (with an average value of about 5) due solely to variability in the sulfur content of the air.

In regions of the globe where the concentrations of sulfur oxides and nitrogen oxides are unusually high, due primarily to the combustion of fossil fuels, rainwater consists of a mixture of sulfuric acid, nitric acid, and water as well as other chemicals. In such circumstances, the pH of rainwater can reach values of 4 or lower. This is the phenomenon of *acid rain,* which can cause damage to fish, soil, crops, and property. It also has important consequences for geochemical cycling of various minerals and their constituent elements.

Exercise 5.6. The concentration of sulfate in the air over the remote oceans is about 1 μg per cubic meter of air. Assuming that this sulfate is sulfuric acid and that in a cloud all of it is dissolved in the cloud drops, what will be the pH of the water in a cloud that contains 0.50 g of liquid water in a cubic meter of air? Ignore the effects of other chemicals.

Solution. H_2SO_4 is a strong acid that is 100% ionized in water

$$H_2SO_4(l) + 2H_2O(l) \rightleftharpoons SO_4^{2-}(aq) + 2H_3O^+(aq)$$

Therefore, 2 moles of $H_3O^+(aq)$ are formed for every 1 mole of $H_2SO_4(l)$ that dissolves in water. Hence, the pH of the water is given by

$$pH = -\log[H_3O^+(aq)] = -\log\{2[H_2SO_4(l)]\}$$

Therefore, to solve the problem we need to find the molar concentrations of H_2SO_4 that dissolves in the cloud water. The number of moles in 1 μg of H_2SO_4 is $10^{-6}/96$, where 96 is the molecular weight of H_2SO_4. Since this amount of H_2SO_4 dissolves in 0.50 g (or 0.50×10^{-3} L) of cloud water

$$[H_2SO_4(l)] = \frac{10^{-6}}{96 \times 0.50 \times 10^{-3}} = 2.0 \times 10^{-5} \text{ M}$$

Therefore,

$$\text{pH of cloud water} = -\log(4.0 \times 10^{-5}) = 4.4$$

This illustrates that even in clean, remote environments the pH of cloud water (and therefore rain) is potentially quite acidic. Of course, some of this acidity will be offset by the absorption of ammonia into cloud water, although the concentration of ammonia is often low over the oceans. On the other hand, as we will see in Chapter 6, additional sulfate may be produced in cloud drops. The likely potential range of variability of the pH of cloud water and rain over the oceans is illustrated by Exercise (5.19).

Exercises

5.7. Answer, interpret, or explain the following in the light of the principles presented in this chapter.

(a) The conjugated bases of strong acids are weak bases; the conjugated bases of weak acids are strong bases.

(b) A solution of $NaNO_2(aq)$ is basic.

(c) It can generally be assumed that a solution with an excess of $Cl^-(aq)$, $SO_4^{2-}(aq)$, $HCO_3^-(aq)$ etc. with respect to $Na^+(aq)$, $K^+(aq)$, $Mg^{2+}(aq)$ etc. is acidic.

(d) Human blood is a complex aqueous solution that is continually replaced, but the pH of the blood of a healthy person does not vary much from 7.4.

(e) A solution has an optimal buffering capacity when conjugate acid–base pairs are present in equal concentrations.

(f) The addition of calcium carbonate ($CaCO_3$) to small lakes can help counteract the effects of acid rain.

(g) Seawater is a buffered solution (with a pH between 8.1 and 8.3). What do you think are its dominant buffers against acids and bases? Where do these buffers originate?

(h) The range of potential pH values for rain is greater over the continents than over the oceans.

5.8.   What are the concentrations of $H^+(aq)$ and $OH^-(aq)$ in an aqueous solution to which 0.05 M of NaOH is added at 25°C? Is NaOH an acid or a base?

5.9.   Calculate the concentrations of $H^+(aq)$ and $OH^-(aq)$ in an aqueous solution to which 1.0 M of acetic acid (CH_3COOH) is added at 25°C. Assume that the acetic acid is 4% ionized.

5.10.   Identify the conjugate acid–base pairs for the forward reaction (indicate by 1) and the reverse reaction (indicate by 2) of the following reactions.
(a) $HSO_4^-(aq) + H_2O(l) \rightleftarrows H_3O^+(aq) + SO_4^{2-}(aq)$
(b) $H_2PO_4^-(aq) + HCl(l) \rightleftarrows H_3PO_4(aq) + Cl^-(aq)$
(c) $NH_4^+(aq) + CH_3COO^-(aq) \rightleftarrows CH_3COOH(aq) + NH_3(aq)$

5.11.   Determine the concentrations of $OH^-(aq)$ and $H^+(aq)$ ions in a 0.50 M solution of ammonia that is 2% ionized. What is the pH of the solution?

5.12.   The pH of a 0.200 M aqueous solution of hydrocyanic acid (HCN) is 5.05 at 25°C. What is the value of the acid-dissociation constant for HCN at 25°C?

5.13.   What is the concentration of protons in a solution with an initial concentration of nitrous acid (HNO_2) of 0.0050 M at 25°C? The acid-dissociation constant of HNO_2 at 25°C is 5.1×10^{-4}.

5.14.   Calculate the concentrations of $H^+(aq)$, $H_2PO_4^-(aq)$, $HPO_4^{2-}(aq)$, and $PO_4^{3-}(aq)$ ions in a 0.020 M aqueous solution of phosphoric acid (H_3PO_4) at 25°C. The successive acid-dissociation constants at 25°C for phosphoric acid are $K_{a1} = 5.9 \times 10^{-3}$, $K_{a2} = 6.2 \times 10^{-8}$, and $K_{a3} = 4.8 \times 10^{-13}$.

5.15.   What is the pH of a 0.0050 M solution of $NH_4C_2H_3O_2$ at 25°C, and what fraction is hydrolyzed? The acid-dissociation constant for $HC_2H_3O_2$ is 1.8×10^{-5} at 25°C and the base-dissociation constant for NH_3 is 1.8×10^{-5} at 25°C.

5.16.   A sample of water contains 40 g L^{-1} of $Na^+(aq)$, 5.2 g L^{-1} of $Ca^{2+}(aq)$, 75.2 g L^{-1} of $Cl^-(aq)$, and 2.0 g L^{-1} of $SO_4^{2-}(aq)$. The only other ions present are $H_3O^+(aq)$ and $OH^-(aq)$. What is the pH of the solution?

5.17.   What is the pH of a buffer solution that contains $NH_4Cl(aq)$ and $NH_3(aq)$ if $[NH_4Cl(aq)] = 2.00[NH_3(aq)]$? The acid-dissociation constant for $NH_4^+(aq)$ is 5.60×10^{-10}.

5.18. One-tenth liter of a buffer solution contains hydrocyanic acid, HCN(aq), and its conjugate base, CN^-(aq). The solution has a pH of 7.9 and contains 7.0×10^{-4} moles of HCN(aq). (a) What is the molar concentration of CN^-(aq) in the buffer solution? (b) By how much would the pH of the solution change if 1.0×10^{-5} moles of $HClO_4$ were added? (c) By how much would the pH of the solution change if 1.0×10^{-5} moles of NaOH were added? (d) What is the total buffering capacity for acids of the original solution? The acid-dissociation constant for HCN is 4.8×10^{-10}.

5.19. Assuming that in oceanic air sulfate concentrations can range from 0.040 to 1.0 μg m^{-3}, and cloud liquid water contents range from 0.10 to 2.5 g m^{-3}, calculate the range of pH values of cloud water over the oceans. Assume that the sulfate is H_2SO_4 and neglect other effects on the acidity. Do you think that the upper pH value you have calculated would be achieved in nature? If not, why?

Notes

1 The term *alkaline* is essentially a synonym for *basic*; it refers to any solution containing appreciable OH^-(aq) or a substance that can form such a solution. Thus, *alkalis* are soluble strong bases, such as NaOH and KOH. *Alkali metal* means any metal of the group Na, K, Li, Rb, and Cs. In general, the oxides of metallic elements are basic, and the oxides of nonmetallic elements are acidic.

2 The proton has a radius of about 10^{-15} m. Other ions that have electrons associated with them have radii of about 10^{-10} m.

3 Those familiar with the structure of atoms would expect from these definitions that acids are materials in which the outer electron orbitals are not completely filled with electrons, and bases are materials with electrons available for sharing. For example, from the Lewis viewpoint, H^+ is an acid because it has an empty orbital that can accept a pair of electrons, and OH^- is a base because it has pairs of electrons available for sharing.

4 The symbol pH was introduced by a Danish chemist, S. Sørensen; p stands for the Danish word for power and H for hydrogen. With a change in sign, pH is the power of ten of the hydrogen ion concentration in moles per liter.

5 Another way of viewing this is that hydrolysis is the reverse of acid dissociation [cf. Reaction (5.20)]. Thus, the weaker the acid [e.g., $H_2C_2H_3O_2$ in Reaction (5.20)] the more difficult it is to remove a proton from it, and the easier it is for its anion or conjugate base [e.g., $C_2H_3O_2^-$(aq) in Reaction (5.20)] to attach a proton from water (i.e., to hydrolyze).

6 In calculations such as this, where several approximations are made, the solutions should be checked by substituting the derived values back into the original equations to see if reasonable equalities are obtained. This is left as an exercise for the reader for this case.

6

Oxidation–reduction reactions

6.1 Some definitions

In Chapter 5 we saw that, in terms of the Brønsted–Lowry theory, acid–base reactions involve proton transfer. Another large and important group of chemical reactions, particularly in aqueous solutions, involves electron transfer; these are referred to as *oxidation–reduction* (or *redox*) reactions. Redox reactions are involved (1) in photosynthesis, which releases oxygen into the Earth's atmosphere; (2) in the combustion of fuels, which is responsible for rising concentrations of atmospheric carbon dioxide; (3) in the formation of acid precipitation; and (4) in many chemical reactions in Earth sediments.

Oxidation refers to a *loss* of electrons, and *reduction* to a *gain* of electrons. For example, an oxidation reaction is

$$Cu(s) \rightarrow Cu^{2+}(aq) + 2e^- \tag{6.1}$$

where the symbol e^- indicates one free electron (which carries one unit of negative charge). A reduction reaction is

$$2Ag^+(aq) + 2e^- \rightarrow 2Ag(s) \tag{6.2}$$

Since electrons cannot be lost or gained overall, oxidation must always be accompanied by reduction. Thus, Eqs. (6.1) and (6.2) together form a redox reaction

$$Cu(s) + 2Ag^+(aq) \rightarrow Cu^{2+}(aq) + 2Ag(s) \tag{6.3}$$

Equation (6.1) is called the *oxidation half-reaction* and Eq. (6.2) the *reduction half-reaction* for the *overall reaction* Eq. (6.3).

If substance A causes the oxidation of substance B, substance A is called the *oxidizing agent* or *oxidant*. Thus, in Eq. (6.3), $Ag^+(aq)$ is the oxidant, because it causes $Cu(s)$ to lose electrons (note that the oxidant is reduced, that is, it gains electrons). Similarly, if a substance A causes the reduction of substance B, substance A is called the *reducing agent* or *reductant*. In Eq. (6.3) Cu(s) is the reductant, because it causes $Ag^+(aq)$ to gain electrons (note that the reductant is oxidized, that is, it loses electrons).[1]

6.2 Oxidation numbers

In order to deal with oxidation–reduction reactions that are more complex than the simple ones discussed so far, we must introduce the concept of *oxidation numbers* (sometimes called *oxidation states* or *valence states*). Oxidation numbers permit us to identify and balance redox reactions and to determine the oxidant and reductant.

Consider the redox reaction in solution of the ferric ion $Fe^{3+}(aq)$ and hydrogen sulfite $HSO_3^-(aq)$

$$2Fe^{3+}(aq) + HSO_3^-(aq) + H_2O(l) \rightarrow$$
$$2Fe^{2+}(aq) + HSO_4^-(aq) + 2H^+(aq) \tag{6.4}$$

The half-reactions are

$$HSO_3^-(aq) + H_2O(l) \rightarrow HSO_4^-(aq) + 2H^+(aq) + 2e^- \tag{6.5}$$

and,

$$2Fe^{3+}(aq) + 2e^- \rightarrow 2Fe^{2+}(aq) \tag{6.6}$$

In this case, it is difficult to say exactly where the two electrons on the right side of Reaction (6.5) come from. Therefore, we need a method for keeping track of the electrons in reactions such as this. This can be done by making some assumptions. It is assumed that the hydrogen atom in $HSO_3^-(aq)$ carries one unit of positive charge, and that each oxygen atom carries two units of negative charge. Since the total charge on $HSO_3^-(aq)$ is one unit of negative charge, the charge on the sulfur atom is four units of positive charge. [Charge on sulfur atom = molecular charge − charge on one hydrogen atom − charge on three oxygen atoms = −1 −1 −3(−2) = +4.] This *fictitious* charge is called the *oxidation number* of sulfur in $HSO_3^-(aq)$.

Following the same procedure, the oxidation number of sulfur in $HSO_4^-(aq)$ is: −1 −1 −4(−2) = +6. Thus, according to this method of bookkeeping, the two electrons on the right side of Reaction (6.5) originate from the sulfur atom, which changes its oxidation number from $+4$ to $+6$ or, stated another way, from changing its oxidation state from S(IV) to S(VI) – read "sulfur four to sulfur six."[2]

The (arbitrary) rules used in assigning oxidation numbers are as follows:

1. The oxidation number of a monoatomic substance is the charge on the atom [e.g., $Cu^+(aq)$ and $S^{2-}(aq)$ have oxidation numbers of $+1$ and -2, respectively].
2. In ionic binary compounds, the oxidation numbers are the charges per ion. [For example, $CdCl_2$ is an ionic compound, as indicated more clearly by $Cd^{2+}(Cl^-)_2$. Thus, the oxidation number of the cadmium ion is $+2$, and the oxidation number of each of the two chloride ions is -1.] The algebraic sum of the oxidation numbers of the atoms in an ion is equal to the charge on the ion (e.g., zero charge for $CdCl_2$).
3. In nonionic (covalent) compounds, the electrons involved in bond formation are shared, more or less equally, by the bonding atoms. However, to assign oxidation numbers, it is assumed that each bonding electron is attached to a particular atom. If these atoms are identical, the bonding electrons are shared equally between the two atoms. If the atoms are different, all of the electrons in the bond are assigned to the atom that has the greater "attraction" for electrons (as indicated by its *electronegativity*). The most electronegative elements, in order of decreasing electronegativity, are F, O, N, and Cl. Nonmetals are more electronegative than metals. A partial list of electronegativities is given in Table 6.1.

These definitions lead to the following rules for assigning oxidation numbers in polyatomic molecules.

1. The oxidation number of all elements in their elementary state or in any self-binding form is zero (e.g., H in H_2, O in O_2, S in S or S_8).
2. The oxidation number of the oxygen atom is -2 in all of its compounds (except peroxides, such as H_2O_2, where it is -1, and in the elementary and self-binding forms of oxygen, where it is zero).
3. The oxidation number of hydrogen is $+1$ in all of its compounds (except those with metals, where it is -1, and in the elementary and self-binding forms of hydrogen, where it is zero).
4. All other oxidation numbers are assigned in such a way as to make the algebraic sum of the oxidation numbers equal to the net charge on the molecule or ion.

Table 6.1. *Partial list of electronegativities*[a]

H 2.1						
Li 0.97	Be 1.5	B 2.0	C 2.5	N 3.1	O 3.5	F 4.1
Na 1.0	Mg 1.2	Al 1.5	Si 1.7	P 2.1	S 2.5	Cl 2.8
K 0.90	Ca 1.0	Ga 1.8	Ge 2.0	As 2.2	Se 2.5	Br 2.7
Rb 0.89	Sr 1.0	In 1.5	Sn 1.72	Sb 1.82	Te 2.0	I 2.2
Cs 0.86	Ba 0.97	Ti 1.4	Pb 1.5	Bi 1.7	Po 1.8	At 1.9

[a] Note that the electronegativities (i.e., the propensity of elements for electrons) increase from left to right along any row and from bottom to top in any column.

It is also useful to note the following. Except when they are combined with oxygen, many elements have only one oxidation number (in addition to zero for the uncombined element), for example, $+1$ for the alkali metals (Na, K, Li, Rb, and Cs), $+2$ for the alkaline earth metals (Ca, Sr, Ba, and Mg), and -1 for the halogens (F, Cl, Br, I, and At). Other elements can have several oxidation numbers.

Note that in Reaction (6.5), the oxidation number of sulfur is increased (from $+4$ to $+6$), and this is the oxidation half-step (i.e., electrons are released by the reaction). In Reaction (6.6), the oxidation number of the ferric ion is decreased (from $+3$ to $+2$), and this is the reduction half-step (i.e., electrons are taken up by the reaction). This observation can be generalized as follows: an increase in oxidation number represents oxidation (the reductant is oxidized) and a decrease in oxidation number represents reduction (the oxidant is reduced).

Exercise 6.1. Assign oxidation numbers to all of the atoms and identify the elements that are oxidized and those that are reduced in the following two reactions.

(a) $2HNO_3(g) + 3H_2S(g) \rightarrow 2NO(g) + 3S(s) + 4H_2O(l)$

(b) $2Cu^{2+}(aq) + 2H_2O(l) \rightarrow 2Cu(s) + O_2(g) + 4H^+(aq)$

Solution.

(a) The oxidation numbers of the atoms on the left side of the reaction are for $HNO_3(g)$: $+1$ for H, -2 for oxygen and therefore $+5$

for N; for $H_2S(g)$: $+1$ for H and therefore -2 for S. For the atoms on the right side of the reaction the oxidation numbers are for $NO(g)$: -2 for oxygen and therefore $+2$ for N; for S the oxidation number is 0; and for $H_2O(l)$: $+1$ for H and -2 for oxygen. Hence, the reaction decreases the oxidation number for nitrogen (from $+5$ to $+2$); therefore nitrogen is reduced. The oxidation number of sulfur, on the other hand, is increased (from -2 to 0); therefore sulfur is oxidized. The oxidation numbers of the hydrogen and oxygen remain unchanged.

(b) The oxidation numbers of the atoms on the left side of the reaction are $+2$ for Cu, $+1$ for H, and -2 for oxygen. On the right side of the reaction they are 0 for Cu, 0 for oxygen, and $+1$ for H. Hence, the reaction decreases the oxidation number of Cu (from $+2$ to 0); therefore copper is reduced. The oxidation number of oxygen is increased (from -2 to 0); therefore oxygen is oxidized.

6.3 Balancing oxidation–reduction reactions

A balanced chemical equation must have the same number and types of atoms on both sides of the equation, and the sum of the electric charges must be the same for the reactants as for the products of the reaction. If all the reactants and products are known, the equation for a redox reaction may be balanced by the *half-reaction method*. (Another method, called the *oxidation-number method*, may also be used, but for our purposes knowledge of one method is sufficient.)

The half-reaction method involves application of the following sequential steps.

Step 1. Write down the overall unbalanced equation for the reaction.

Step 2. Write down the unbalanced equations for the oxidation half-reaction and for the reduction half-reaction (species should not be written as free atoms or ions unless they exist in these forms).

Step 3. For each of the half-reactions, first balance the atoms that undergo oxidation and reduction. Then balance atoms other than oxygen and hydrogen. Finally, balance the oxygen and hydrogen atoms. [In neutral or acidic solutions, H_2O and $H^+(aq)$ may be added to balance the oxygen and hydrogen atoms.[3] The oxygen atoms are balanced first. For each excess oxygen atom on one side of the equation, balance is achieved by adding one H_2O to the other side. Then $H^+(aq)$ is used to balance the hydrogen atoms. O_2 and H_2 are not used to balance the oxygen and hydrogen atoms unless they are known to participate in the reaction. For basic solutions, $OH^-(aq)$ can be used to balance the half-reactions. For each excess oxygen atom on one side of the equation, balance is achieved by adding one H_2O to the other side. Hydrogen balance is achieved by adding one OH^- for each excess hydrogen atom on the *same* side of the equation as the excess exists, and one H_2O on the *other* side of the equation. If both oxygen and hydrogen atoms are in excess on the same side of the equation, add an OH^- on the other side of the equation for each pair of oxygen and hydrogen atoms that are in excess.]

Step 4. For each of the half-reactions, balance the electric charge by adding electrons to the right side of the oxidation half-reaction and to the left side of the reduction half-reaction.

Step 5. Multiply each half-reaction by a number that makes the total number of electrons lost by the reductant equal to the number of electrons gained by the oxidant.

Step 6. Add the two half-reactions to get the overall reaction. (Cancel any terms that are identical on both sides of the reaction; all electrons should cancel.)

Step 7. Check the overall reaction for conservation of the atoms of each element and net charge.

The following two exercises illustrate the application of this method, first to an acidic reaction and then to a basic reaction.

Exercise 6.2. Balance the equation for the following redox reaction in an aqueous solution

$$HNO_3(aq) + H_2S(aq) \rightarrow NO(g) + S(s) + H_2O(l)$$

Solution.

Step 1. The unbalanced equation is given above, which, for an aqueous solution, may also be written

$$H^+(aq) + NO_3^-(aq) + H_2S(aq) \rightarrow NO(g) + S(s) + H_2O(l) \qquad (6.7)$$

Step 2. The oxidant is $NO_3^-(aq)$, because it contains N, which undergoes a decrease in oxidation number [from $+5$ on the right side of Reaction (6.7) to $+2$ on the left side]. Therefore, the unbalanced reduction half-reaction is

$$NO_3^-(aq) \rightarrow NO(g) \qquad (6.7a)$$

The reductant in Reaction (6.7) is $H_2S(aq)$, because it contains $S(s)$, which undergoes an increase in oxidation number [from -2 on the right side of Reaction (6.7) to zero on the left side]. Therefore, the unbalanced oxidation half-reaction is

$$H_2S(aq) \rightarrow S(s) \qquad (6.7b)$$

Step 3. We must now balance the oxygen atoms in Reaction (6.7a). Since this is an acidic solution, and there are two excess oxygen atoms on the right side of Reaction (6.7a), we do this by first adding $2H_2O(l)$ to the right side of Reaction (6.7a)

$$NO_3^-(aq) \rightarrow NO(g) + 2H_2O(l)$$

Since there are now four excess hydrogen atoms on the right side of the last reaction, we add $4H^+(aq)$ to the left side to obtain the balanced reduction half-reaction

$$NO_3^-(aq) + 4H^+(aq) \rightarrow NO(g) + 2H_2O(l) \qquad (6.7c)$$

To balance the oxidation half-reaction (6.7b), since there are two excess hydrogen atoms on the left side, we add $2H^+(aq)$ to the right side

$$H_2S(aq) \rightarrow S(s) + 2H^+(aq) \qquad (6.7d)$$

Step 4. In Reaction (6.7c) the net charge on the left side is $+3$, and on the right side it is 0. Therefore, we add three electrons to the left side

$$NO_3^-(aq) + 4H^+(aq) + 3e^- \rightarrow NO(g) + 2H_2O(l) \qquad (6.7e)$$

In Reaction (6.7d) the net charge on the left is 0 and on the right it is $+2$. Therefore, two electrons must be added to the right side

$$H_2S(aq) \rightarrow S(s) + 2H^+(aq) + 2e^- \qquad (6.7f)$$

Step 5. To equalize the electrons lost by the reductant and gained by the oxidant, we must multiply Reaction (6.7e) by 2 and Reaction (6.7f) by 3 to give

$$8H^+(aq) + 2NO_3^-(aq) + 6e^- \rightarrow 2NO(g) + 4H_2O(l) \qquad (6.7g)$$

and,

$$3H_2S(aq) \rightarrow 3S(s) + 6H^+(aq) + 6c^- \qquad (6.7h)$$

[Note that in Reaction (6.7h) the oxidation number of each of the three sulfur atoms increases by two (from -2 in H_2S to zero in S), for a total increase of six, which is consistent with the six electrons on the right side of reaction (6.7h). Similarly, in Reaction (6.7g) the oxidation number of each of the two nitrogen atoms decreases by three (from $+5$ in NO_3^- to $+2$ in NO), for a total decrease of six, which is consistent with the six electrons on the left side of Reaction (6.7g). This is a useful check on the half-reactions.]

Step 6. Adding Reactions (6.7g) and (6.7h) yields

$$8H^+(aq) + 2NO_3^-(aq) + 3H_2S(aq) + 6e^- \rightarrow$$
$$2NO(g) + 4H_2O(l) + 3S(s) + 6H^+(aq) + 6e^-$$

Canceling $6H^+(aq)$ and $6e^-$ from both sides of this reaction gives

$$2H^+(aq) + 2NO_3^-(aq) + 3H_2S(aq) \rightarrow 2NO(g) + 4H_2O(l) + 3S(s) \quad (6.8)$$

Step 7. Inspection shows that Reaction (6.8) conserves the various atoms and electric charge.

Exercise 6.3. Balance the equation for the following redox reaction in an aqueous solution

$$Cr(OH)_3(s) + OCl^-(aq) + OH^-(aq) \rightarrow$$
$$CrO_4^{2-}(aq) + Cl^-(aq) + H_2O(l) \quad (6.9)$$

Solution.

Step 1. The unbalanced equation is given by Reaction (6.9).

Step 2. The oxidant is $OCl^-(aq)$, because it contains Cl, which undergoes a decrease in oxidation number [from $+1$ on the right side of Reaction (6.9) to -1 on the left side]. Therefore, the unbalanced reduction half-reaction is

$$OCl^-(aq) \rightarrow Cl^-(aq) \quad (6.9a)$$

The reductant in Reaction (6.9) is $Cr(OH)_3(s)$, because it contains Cr which undergoes an increase in oxidation number [from $+3$ on the right side of Reaction (6.9) to $+6$ on the left side]. Therefore, the unbalanced oxidation half-reaction is

$$Cr(OH)_3(s) \rightarrow CrO_4^{2-}(aq) \quad (6.9b)$$

Step 3. We must now balance the oxygen atoms in Reaction (6.9a). Since this is a basic solution, and there is one excess oxygen atom on the left side of Reaction (6.9a), we first add one $H_2O(l)$ to the right side of Reaction (6.9a)

$$OCl^-(aq) \rightarrow Cl^-(aq) + H_2O(l)$$

Since there are now two excess hydrogen atoms on the right side of the last equation, we add two OH^-(aq) to the right side and two H_2O(l) to the left side

$$OCl^-(aq) + 2H_2O(l) \rightarrow Cl^-(aq) + H_2O(l) + 2OH^-(aq)$$

or, on canceling one H_2O(l) from each side

$$OCl^-(aq) + H_2O(l) \rightarrow Cl^-(aq) + 2OH^-(aq) \qquad (6.9c)$$

Since there is one excess oxygen atom on the right side of Reaction (6.9b), as a first step in achieving atomic balance, we add one H_2O(l) to the left side

$$Cr(OH)_3(s) + H_2O(l) \rightarrow CrO_4^{2-}(aq)$$

Since there are now five excess hydrogen atoms on the left side of the last reaction, to achieve hydrogen balance we add five OH^-(aq) to the left side and five H_2O(l) to the right side

$$Cr(OH)_3(s) + H_2O(l) + 5OH^-(aq) \rightarrow CrO_4^{2-}(aq) + 5H_2O(l)$$

or, on canceling one H_2O(l) from each side

$$Cr(OH)_3(s) + 5OH^-(aq) \rightarrow CrO_4^{2-}(aq) + 4H_2O(l) \qquad (6.9d)$$

Step 4. In Reaction (6.9c) the net charge on the left side is 0 and on the right side it is -2. Therefore, we add two electrons to the left side

$$OCl^-(aq) + H_2O(l) + 2e^- \rightarrow Cl^-(aq) + 2OH^-(aq) \qquad (6.9e)$$

In Reaction (6.9d) the net charge on the left side is -5 and on the right side it is -2. Therefore, we add three electrons to the right side

$$Cr(OH)_3(s) + 5OH^-(aq) \rightarrow CrO_4^{2-}(aq) + 4H_2O(l) + 3e^- \qquad (6.9f)$$

Step 5. To equalize the electrons lost by the reductant and gained by the oxidant, we must multiply Reaction (6.9e) by 3 and Reaction (6.9f) by 2, to give

$$3OCl^-(aq) + 3H_2O(l) + 6e^- \rightarrow 3Cl^-(aq) + 6OH^-(aq) \qquad (6.9g)$$

and,

$$2Cr(OH)_3(s) + 10\ OH^-(aq) \rightarrow 2CrO_4^{2-}(aq) + 8H_2O(l) + 6e^- \qquad (6.9h)$$

[The reader should check Reactions (6.9g) and (6.9h) for consistency between electron losses or gains and changes in oxidation numbers – see step 5 in Exercise 6.2.]

Step 6. Adding Reactions (6.9g) and (6.9h) yields

$$3OCl^-(aq) + 3H_2O(l) + 2Cr(OH)_3(s) + 10\{OH^-(aq)\} + 6e^- \rightarrow$$
$$3Cl^-(aq) + 6OH^-(aq) + 2CrO_4^{2-}(aq) + 8H_2O(l) + 6e^-$$

Canceling $3H_2O(l)$, $6OH^-(aq)$, and $6e^-$ from both sides of the equation gives

$$2Cr(OH)_3(s) + 3OCl^-(aq) + 4OH^-(aq) \rightarrow$$
$$2CrO_4^{2-}(aq) + 3Cl^-(aq) + 5H_2O(l) \qquad (6.10)$$

Step 7. Inspection shows that Reaction (6.10) conserves the various atoms and electric charge.

Photosynthesis is probably the most important redox reaction. During photosynthesis in green plants, light energy is used to convert CO_2 and H_2O into oxygen and energy-rich organic compounds called carbohydrates (e.g., glucose, $C_6H_{12}O_6$). Without photosynthesis the Earth's atmosphere would have no oxygen. If photosynthesis stopped, most living things would die in a few years. The reverse of photosynthesis is called respiration. This occurs primarily at night when carbohydrates in plants react with oxygen to release CO_2 and H_2O back into the atmosphere. In prehistoric times green plants, growing in warm climates in an atmosphere much richer in CO_2 than the present atmosphere, "tied up" vast quantities of carbon in the form of glucose. As a result, enormous deposits of fossil fuels (i.e., coal, oil, and gas) were deposited in the ground. The combustion of these fuels in modern times is releasing this carbon back into the atmosphere in the form of CO_2 gas. Carbon dioxide is a "greenhouse gas" that contributes to global warming.

Exercise 6.4. The unbalanced equation for photosynthesis is

$$CO_2(g) + H_2O(l) \rightarrow C_6H_{12}O_6(s) + O_2(g) \qquad (6.11)$$

Balance the equation for this reaction.

Solution.

Step 1. The unbalanced reaction is (6.11).

Step 2. The oxidant is $CO_2(g)$ because it contains carbon that undergoes a decrease in oxidation number [from $+4$ in $CO_2(g)$ to 0 in $C_6H_{12}O_6$]. Therefore, the unbalanced reduction half-reaction is

$$CO_2(g) \rightarrow C_6H_{12}O_6(s) \qquad (6.11a)$$

The reductant is $H_2O(l)$, because it contains oxygen which undergoes an increase in oxidation number [from -2 in $H_2O(l)$ to 0 in $O_2(g)$]. The unbalanced oxidation half-reaction is

$$H_2O(l) \rightarrow O_2(g) \qquad (6.11b)$$

Step 3. We must first balance the carbon in Reaction (6.11a) by multiplying the left side by six

$$6CO_2(g) \rightarrow C_6H_{12}O_6(s)$$

Since the overall reaction is neutral, and there are six excess oxygen atoms on the left side, we must add six $H_2O(l)$ to the right side

$$6CO_2(g) \rightarrow C_6H_{12}O_6(s) + 6H_2O(l)$$

Since there are now 24 excess hydrogen atoms on the right side, we must add 24 $H^+(aq)$ ions to the left side to obtain the balanced reduction half-reaction

$$6CO_2(g) + 24H^+(aq) \rightarrow C_6H_{12}O_6(s) + 6H_2O(l) \qquad (6.11c)$$

To balance Reaction (6.11b), since there is one excess oxygen atom on the right side, we must first add one $H_2O(l)$ on the left side

$$2H_2O(l) \rightarrow O_2(g)$$

There are now four excess hydrogen atoms on the left side, which can be balanced by adding four $H^+(aq)$ to the right side

$$2H_2O(l) \rightarrow O_2(g) + 4H^+(aq) \qquad (6.11d)$$

Step 4. To balance Reaction (6.11c) electrically

$$6CO_2(g) + 24H^+(aq) + 24e^- \rightarrow C_6H_{12}O_6(s) + 6H_2O(l) \qquad (6.11e)$$

To balance Reaction (6.11d) electrically

$$2H_2O(l) \rightarrow O_2(g) + 4H^+(aq) + 4e^- \qquad (6.11f)$$

Step 5. To make the number of electrons gained in Reaction (6.11e) equal to the number released in Reaction (6.11f), we multiply Reaction (6.11f) by six

$$12H_2O(l) \rightarrow 6O_2(g) + 24H^+(aq) + 24e^- \qquad (6.11g)$$

Step 6. Adding Reactions (6.11e) and (6.11g) and canceling terms yields

$$6CO_2(g) + 6H_2O(l) \rightarrow C_6H_{12}O_6(s) + 6O_2(g) \qquad (6.12)$$

Step 7. Inspection shows that Reaction (6.12) conserves the various atoms and electric charge.

6.4 Half-reactions in electrochemical cells

We have seen that in a redox reaction electrons are transferred from one species to another. In an *electrochemical* (or *galvanic* or *voltaic*) *cell* this transfer takes place along a wire and therefore generates an electric current. For example, for the silver–copper reaction considered at the beginning of this chapter the galvanic cell is shown in Figure 6.1. It consists of a piece of silver metal in a beaker containing a solution of $AgNO_3$, and a piece of copper in another beaker containing $CuSO_4$ solution. The two pieces of metal (called *electrodes*) are connected externally by a wire through an ammeter (to measure the flow of current). The electrodes are connected internally by the electrically conducting aqueous solution (called the *electrolyte*), in which the electrodes are immersed, and by a solution of $NaNO_3$ (called a salt bridge). In this circuit it is observed that the copper electrode slowly dissolves, the silver electrode gains mass, and electrons flow through the wire from the copper to the silver electrode. This is because at the copper electrode the oxidation half-reaction (6.1) takes place

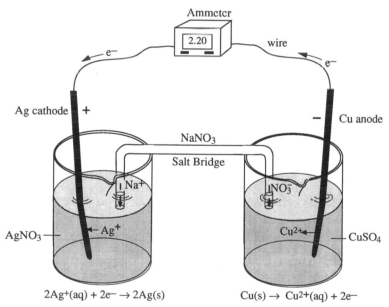

$$2Ag^+(aq) + 2e^- \rightarrow 2Ag(s) \qquad Cu(s) \rightarrow Cu^{2+}(aq) + 2e^-$$

Figure 6.1. An electrochemical (or galvanic) cell. The reduction half-reaction occurs at the silver electrode, which is therefore the cathode and is positively charged. The oxidation half-reaction occurs at the copper electrode, which is therefore the anode and is negatively charged.

$$CuS \rightarrow Cu^{2+}(aq) + 2e^- \tag{6.1}$$

and at the silver electrode the reduction half-reaction (6.2) takes place

$$2Ag^+(aq) + 2e^- \rightarrow 2Ag(s) \tag{6.2}$$

Thus, the beaker on the right side of Figure 6.1 is called the *oxidation half-cell,* and the beaker on the left side the *reduction half-cell.*

The electrode at which the oxidation half-reaction takes place (i.e., the copper electrode in Fig. 6.1) is called the *anode.* Electrons flow from the anode through the wire; therefore, the anode is considered to be the negatively charged electrode. The electrode at which the reduction half-reaction occurs (i.e., the silver electrode in Fig. 6.1) is called the *cathode.* Since electrons flow toward the cathode through the wire, it is the positively charged electrode. Within the cell itself, negatively charged ions (called *anions*) drift toward the anode in order to neutralize the positive ions released into the solution by the oxidation half-reaction (see Fig. 6.1).[4] Conversely, positively charged ions (called *cations*) drift toward the cathode to keep that half-cell neutral (see Fig. 6.1).

6.5 Strengths of oxidants and reductants; standard cell and half-cell potentials

The two half-reactions of any redox reactions can be thought of as occurring in the two half-cells of an electrochemical cell. We will now show how this provides a quantitative method for comparing the strengths of various oxidants and reductants and the spontaneous direction in which a redox reaction will occur.

If the ammeter in Figure 6.1 were replaced by a voltmeter, we could measure the electric potential difference (in volts, indicated by V) between the two electrodes of an electrochemical cell. Experiments show that for any two metal electrodes (e.g., Cu and Ag), this potential difference depends on the relative concentrations of $Cu^{2+}(aq)$ and $Ag^+(aq)$ in the two solutions, as well as temperature, pressure, etc. However, if the temperature is kept at 25°C, the pressure is constant at 1 atm, and the concentrations of the two aqueous ions are kept equal (say at 1 M), then, provided not too much current is drawn, any two metal electrodes generate a steady potential difference the magnitude of which depends on the nature of the electrodes (e.g., 0.46 V when the electrodes are Cu and Ag).

We could set up a number of electrochemical cells, each with a different pair of metal electrodes in contact with solutions of their metal ions, and measure the steady potential difference generated by each cell under *standard conditions* (i.e., 25°C, 1 atm and 1 M concentrations of the metal ions).[5] The voltages obtained in this way are called *standard cell potentials* (E^0_{cell}). The greater the tendency (or driving force) for a redox reaction to occur (with reactants and products in their standard states), the greater will be its standard cell potential.

Exercise 6.5. An electrochemical cell has a standard cell potential of 2.0 V. How much electrical work can it do in 1.0 minute if it operates under standard conditions and a steady current of 1.0 ampere is drawn from the cell? If the cell is that shown in Figure 6.1, how many moles of $Ag^+(aq)$ ions are neutralized by electrons in 1.0 minute? [The electric charge on 1 mole of electrons, called the *Faraday constant* (F), is about 96,489 coulombs.][6]

Solution. When a current I (in amperes) flows through a potential difference E (in volts) for time t (in seconds), the electrical work done is, from elementary electrical theory, EIt (in joules). Therefore, since the cell provides a potential difference of 2.0 V (assuming no potential loss at the electrodes), the electrical work done by the cell in 1 minute, when a current of 1.0 ampere flows, is $2.0 \times 1.0 \times 60 = 120$ J.

Since electric charge (in coulombs) is given by the product of current (in amperes) and time (in seconds), the electric charge that flows through the circuit in time t is $It = yF$, where y is the number of moles of electrons that pass in t seconds. Therefore, $It - 96,489y$ or, since $It - 1.0 \times 60$ ampere-sec, $y = 6.2 \times 10^{-4}$ moles of electrons in 1 minute. In the half–cell depicted on the left side of Figure 6.1, the electrons neutralize the $Ag^+(aq)$ ions in solution. Also, it can be seen from Reaction (6.1) that 1 mole of electrons reacts with 1 mole of $Ag^+(aq)$ ions. Hence, the number of moles of $Ag^+(aq)$ ions that are neutralized in 1 minute is 6.2×10^{-4}.

So far we have considered only the magnitude of the standard cell potential E^0_{cell}. By convention, E^0_{cell} is considered to have a positive value if the reaction proceeds spontaneously in the forward direction. Thus, Reaction (6.3)

$$Cu(s) + 2Ag^+(aq) \rightarrow Cu^{2+}(aq) + 2Ag(s) \qquad (6.3)$$

proceeds spontaneously in the forward direction (i.e., from left to right); therefore its standard cell potential is 0.46 V. Conversely, the reaction

$$Cu^{2+}(aq) + 2Ag(s) \rightarrow Cu(s) + 2Ag^+(aq)$$

proceeds spontaneously from right to left; therefore, its standard cell potential is -0.46 V.

The standard cell potential depends on the electronic propensity of the species involved in the overall reaction. Thus, the value of E^0_{cell} for Reaction (6.3) depends on the tendency of Cu(s) to give up electrons [see half-reaction (6.1)] and the tendency of $Ag^+(aq)$ to gain electrons [see half-reaction (6.2)]. If we could define *standard half-cell* (or *electrode*) *potentials* [e.g., for the half-reaction (6.1) and for the half-reaction (6.2)], they would provide measures of the strengths of various oxidants and reductants. This can be done by (arbitrarily) assigning one specific half-reaction a standard half-cell potential of 0. The half-reaction chosen for this purpose is that for the hydrogen half-cell

$$H_2(g) \rightarrow 2H^+(aq) + 2e^- \qquad (6.13)$$

for which $E^0_{cell} = 0$ (for the forward or reverse reaction) when the reactants and products are in their standard states.

Standard half-cell potentials can be measured for other half-reactions by measuring the magnitude of the standard cell potential (E^0_{cell}) when each half-cell is combined with the hydrogen half-cell (we need not concern ourselves here with the technical aspects of such measurements). For example, when the $Cu-Cu^{2+}$ half-cell [see Reaction (6.1)] is combined with the hydrogen half-cell, the standard cell potential is 0.34 V and electrons flow in the external (wire) part of the circuit *from* the hydrogen electrode *to* the copper electrode. Comparing this with the situation shown in Figure 6.1, we see that the silver electrode has been replaced by a hydrogen electrode (and $AgNO_3$ by H_2SO_4), and the flow of electrons through the wire has been reversed. Therefore, in place of Reaction (6.1), (6.13) is the oxidation half-cell reaction. The reduction half-cell reaction [which replaces reaction (6.2)] is the reverse of Reaction (6.1), that is,

$$Cu^{2+}(aq) + 2e^- \rightarrow Cu(s) \qquad (6.14)$$

From the addition of Reactions (6.13) and (6.14), we can see that the spontaneous overall cell reaction is now

$$H_2(g) + Cu^{2+}(aq) \rightarrow 2H^+(aq) + Cu(s) \qquad (6.15)$$

for which $E^0_{cell} = 0.34$ V. Since, by definition, Reaction (6.13) does not generate any electric potential difference, the absolute magnitude of the standard half-cell (or electrode) potential for Reaction (6.14) is $E_{red}^0 = 0.34$ V.

If zinc is made the electrode of one half-cell and hydrogen is the other half-cell, the standard cell potential is found by measurement to be 0.76 V, and electrons flow in the wire *from* the zinc electrode *to* the hydrogen electrode. Comparing this with the situation shown in Figure 6.1, we see that the silver electrode has been replaced by a hydrogen electrode (and AgNO$_3$ by H$_2$SO$_4$) and the copper electrode by a zinc electrode [and CuSO$_4$ by Zn(NO$_3$)$_2$]. Therefore, in place of Reaction (6.1), we now have for the oxidation half-cell reaction

$$Zn(s) \rightarrow Zn^{2+}(aq) + 2e^- \qquad (6.16)$$

and, for the reduction half-cell reaction [replacing Reaction (6.2)]

$$2H^+(aq) + 2e^- \rightarrow H_2(g) \qquad (6.17)$$

Therefore, the spontaneous overall cell reaction is

$$Zn(s) + 2H^+(aq) \rightarrow Zn^{2+}(aq) + H_2(g) \qquad (6.18)$$

for which $E_{cell}^0 = 0.76$ V. Since, by definition, Reaction (6.17) does not generate any potential, the magnitude of the standard half-cell (or electrode) potential for Reaction (6.16) is $E_{ox}^0 = 0.76$ V.

The question now arises as to the sign to be attached to the magnitudes of the electrode potentials for copper and zinc derived above. Clearly, they should be given opposite signs, because in the Cu–H$_2$ cell the electrons in the wire move from the hydrogen electrode to the copper electrode; whereas, in the Zn–H$_2$ cell they move from the zinc to the hydrogen electrode. Whether the negative sign is attached to the copper or to the zinc electrode potential is, of course, a matter of convention. The convention that has been adopted is that *if an electrode forms part of the half-cell in which the reduction reaction takes place when it is coupled with a hydrogen half-cell, the electrode potential is assigned a positive value.* Conversely, *if an electrode forms part of the half-cell in which the oxidation reaction takes place when it is coupled with a hydrogen half-cell, the electrode potential is assigned a negative value.* Applying this convention to the Cu–H$_2$ cell, we see from Reaction (6.14) that Cu has a positive electrode potential (i.e., $E_{red}^0 = 0.34$ V); applying it to the Zn–H$_2$ cell, we see from Reaction (6.16) that zinc has a negative electrode potential (i.e., $E_{ox}^0 = -0.76$ V). If we write the oxidation half-cell reaction (6.16) in the form of its reverse reduction half-reaction

$$Zn^{2+}(aq) + 2e^- \rightarrow Zn(s)$$

the negative sign associated with the zinc electrode potential (-0.76 V) can be interpreted as indicating that the spontaneous reaction proceeds from right to left. Similarly, the fact that copper has a positive electrode potential (0.34 V) indicates that the reduction reaction (6.14) proceeds spontaneously from left to right.

The greater the magnitude (with sign) of an electrode potential, the greater is the driving force for the reduction half-reaction to take place in that half-cell. For example, Zn and Cu have electrode potentials of -0.76 V and 0.34 V, respectively. Therefore, when paired together, Cu will be involved in the reduction half-reaction; that is, it will be the oxidant. Zinc will be involved in the oxidation half-reaction, that is, it will be the reductant.

With the above sign convention, we can write

$$E^0_{cell} = E^0_{ox} + E^0_{red} \tag{6.19}$$

where the magnitude and sign of E^0_{red} are the same as that for the electrode potential of the reduction half-reaction, and E^0_{ox} has the same magnitude but the *opposite* sign as the electrode potential it would have were it to serve as a reduction half-reaction. If E^0_{cell} is positive, the overall or net chemical reaction (obtained by adding the two half-reactions) is spontaneous from left to right, with a driving force that is proportional to the magnitude of E^0_{cell}. The following exercise should make these points clear.

Exercise 6.6. An electrochemical cell has electrodes made of zinc and copper and operates under standard conditions. Which electrode is the anode and which the cathode? Which way will electrons flow in the external (wire) portion of the circuit? What is the maximum electric potential difference that this cell can generate? Will Zn(s) spontaneously reduce $Cu^{2+}(aq)$, or will Cu(s) spontaneously reduce $Zn^{2+}(aq)$?

Solution. As we have just seen, the electrode potentials of copper and zinc are 0.34 V and -0.76 V, respectively. By convention, the greater the value of the electrode potential the more likely it is that the reduction half-reaction will take place at that electrode. Therefore, the reduction half-reaction takes place at the copper electrode and the oxidation half-reaction at the zinc electrode. Hence, by definition, the zinc electrode is the anode and the copper electrode the cathode.

In the wire portion of the circuit, the electrons always flow from the anode (where electrons are released by the oxidation reaction) to the cathode (where electrons are taken up by the reduction reaction). Therefore, in this case, the electrons flow from the zinc electrode to the copper electrode. (This is consistent with the fact that *negatively* charged particles always flow *up* a potential gradient, that is, from the electrode with the lower potential, zinc in this case, to the electrode with the higher potential, copper in this case.)

The maximum electric potential difference that the cell can generate, assuming no inefficiencies, is E^0_{cell}. The half-reactions and their electric potentials are

$$Zn(s) \rightarrow Zn^{2+}(aq) + 2e^- \qquad E^0_{cell} = -(-0.76\ V) = 0.76\ V$$
$$Cu^{2+}(aq) + 2e^- \rightarrow Cu(s) \qquad E^0_{red} = 0.34\ V$$

Net: $Zn(s) + Cu^{2+}(aq) \rightarrow Zn^{2+}(aq) + Cu(s) \quad E^0_{cell} = 1.10\ V$

Since E^0_{cell} is positive the net reaction is spontaneous from left to right. Therefore, $Zn(s)$ reduces $Cu^{2+}(aq)$.

Shown in Table 6.2 are standard electrode potentials for some half-cell reactions. Note that in Table 6.2:

- All of the reactions are listed in the form of reduction half-reactions. (The corresponding oxidation half-reactions are in the opposite directions to those shown in Table 6.2.)
- A positive value of E^0 indicates that the spontaneous reaction proceeds from left to right (i.e., reduction is the spontaneous reaction). A negative value of E^0 indicates that the spontaneous reaction proceeds from right to left (i.e., oxidation is the spontaneous reaction).

A redox reaction must involve an oxidation half-reaction combined with a reduction half-reaction. Therefore, one reactant in an overall redox reaction must come from the left side of a table of reduction half-reactions (i.e., the reduced form in the half-equation) and one reactant must come from the right side of the table (i.e., the oxidized form in the half-equation). Such combinations could be of two types: (a) those for which the oxidized form lies *below* the reduced form (e.g., $O_3(g)$ and $Ag(s)$ – see Table 6.2), and (b) those for which the oxidized form lies *above* the reduced form (e.g., $Fe^{2+}(aq)$ and $H_2(g)$ – see Table 6.2). Under standard conditions, type (a) combinations produce significant redox reactions in the forward direction (because $E^0_{cell} = E^0_{ox} + E^0_{red}$ is positive).[7] For type (b) combinations the redox reaction is not significant in the forward direction under standard conditions (because E^0_{cell} is negative). Stated in another way: *Under standard conditions the reduced form of any couple [e.g., Li(s) for the*

Basic physical chemistry

Table 6.2. *Some standard electrode (or half-cell) potentials*[a]

	Half-reaction	E^0 (volt)	Acidic or basic solution	
VERY STRONG REDUCING AGENTS	$Li^+(aq) + e^- \rightarrow Li(s)$	-3.045	Acidic	VERY WEAK OXIDIZING AGENTS
	$Na^+(aq) + e^- \rightarrow Na(s)$	-2.714	Acidic	
	$Mg^{2+}(aq) + 2e^- \rightarrow Mg(s)$	-2.73	Acidic	
	$Al^{3+}(aq) + 3e^- \rightarrow Al(s)$	-1.66	Acidic	
	$Mn^{2+}(aq) + 2e^- \rightarrow Mn(s)$	-1.18	Acidic	
	$SO_4^{2-}(aq) + H_2O(l) + 2e^- \rightarrow SO_3^{2-}(aq) + 2OH^-(aq)$	-0.93	Basic	
	$2H_2O(l) + 2e^- \rightarrow H_2(g) + 2OH^-(aq)$	-0.828	Basic	
	$Zn^{2+}(aq) + 2e^- \rightarrow Zn(s)$	-0.763	Acidic	
	$Cr^{3+}(aq) + 3e^- \rightarrow Cr(s)$	-0.74	Acidic	
	$S(s) + 2e^- \rightarrow S^{2-}(aq)$	-0.48	Basic	
	$Fe^{2+}(aq) + 2e^- \rightarrow Fe(s)$	-0.44	Acidic	
	$Cd^{2+}(aq) + 2e^- \rightarrow Cd(s)$	-0.402	Acidic	
	$PbSO_4(s) + 2e^- \rightarrow Pb(s) + SO_4^{2-}(aq)$	-0.356	Acidic	
	$Co^{2+}(aq) + 2e^- \rightarrow Co(s)$	-0.277	Acidic	
	$Ni^{2+}(aq) + 2e^- \rightarrow Ni(s)$	-0.250	Acidic	
	$Sn^{2+}(aq) + 2e^- \rightarrow Sn(s)$	-0.136	Acidic	
	$2H^+(aq) + 2e^- \rightarrow H_2(g)$	0 (definition)	Acidic	
	$S(s) + 2H^+(aq) + 2e^- \rightarrow H_2S(aq)$	0.14	Acidic	
	$SO_4^{2-}(aq) + 4H^+(aq) + 2e^- \rightarrow 2H_2O(l) + SO_2(g)$	0.17	Acidic	
	$Cu^{2+}(aq) + 2e^- \rightarrow Cu(s)$	0.34	Acidic	
	$O_2(g) + 2H_2O(l) + 4e^- \rightarrow 4OH^-(aq)$	0.40	Basic	
	$H_2SO_3(aq) + 4H^+ + 4e^- \rightarrow S(s) + 3H_2O(l)$	0.45	Acidic	
	$I_2(s) + 2e^- \rightarrow 2I^-(aq)$	0.536	Acidic	
	$O_2(g) + 2H^+(aq) + 2e^- \rightarrow H_2O_2(aq)$	0.682	Acidic	
	$Fe^{3+}(aq) + e^- \rightarrow Fe^{2+}(aq)$	0.771	Acidic	
	$Ag^+(aq) + e^- \rightarrow Ag(s)$	0.799	Acidic	
	$NO_3^-(aq) + 4H^+(aq) + 3e^- \rightarrow NO(g) + 2H_2O(l)$	0.96	Acidic	

Table 6.2. *(Cont.)*

Half-reaction	E^0 (volt)	Acidic or basic solution
$O_2(g) + 4H^+(aq) + 4e^- \rightarrow 2H_2O(l)$	1.229	Acidic
$O_3(g) + H_2O(l) + 2e^- \rightarrow O_2(g) + 2OH^-(aq)$	1.24	Basic
$Cr_2O_7^{2-}(aq) + 14H^+(aq) + 6e^- \rightarrow 2Cr^{3+}(aq) + 7H_2O(l)$	1.33	Acidic
$Cl_2(g) + 2e^{-1} \rightarrow 2Cl^-(aq)$	1.36	Acidic
$Mn_4^-(aq) + 8H^+(aq) + 5e^- \rightarrow Mn^{2+}(aq) + 4H_2O(l)$	1.51	Acidic
$Ce^{4+}(aq) + e^- \rightarrow Ce^{3+}(aq)$	1.61	Acidic
$PbO_2(s) + SO_4^{2-}(aq) + 4H^+(aq) + 2e^- \rightarrow PbSO_4(s) + 2H_2O(l)$	1.685	Acidic
$H_2O_2(aq) + 2H^+(aq) + 2e^- \rightarrow H_2O(l)$	1.766	Acidic
$S_2O_8^{2-}(aq) + 2e^- \rightarrow 2SO_4^{2-}(aq)$	2.01	Acidic
$O_3(g) + 2H^+(aq) + 2e^- \rightarrow O_2(g) + H_2O(l)$	2.07	Acidic
$F_2(s) + 2e^- \rightarrow 2F^-(aq)$	2.87	Acidic

VERY WEAK REDUCING AGENTS

VERY STRONG OXIDIZING AGENTS

[a]The half-reactions are expressed as reductions. The more propensity a species couple has for supplying electrons, that is, the better its reducing power, the more negative is its E^0 value [e.g., the couple $Li^+(aq) - Li(s)$ has a superior reducing capability to the couple $Na^+(aq) - Na(s)$]. Thus, the better the reducing capability of a couple, the higher up it occurs in this table. Conversely, the more propensity a couple has for consuming electrons, that is, the better its oxidizing power, the more positive is its E^0 value listed in this table. In other words, large negative values of E^0 indicate that the preferred direction for the half-reaction is from right to left, and large positive values of E^0 indicate that the preferred direction is from left to right. If a half-reaction is reversed, to form an oxidation half-reaction, the sign of E^0 changes. With this sign convention, any two half-reactions that combine to form an overall reaction that is spontaneous from left to right have a positive. (E^0_{cell} $E^0_{cell} = E^0_{ox} + E^0_{red}$).

$L^+(aq) - Li(s)$ couple, or $O_2(g)$ for the $O_3(g) - O_2(g)$ couple] will react to a significant extent (i.e., spontaneously) with the oxidized form of any couple [e.g., $Fe^{2+}(aq)$ for the $Fe^{2+}(aq) - Fe(s)$ couple, or $Ag^+(aq)$ for the $Ag^+(aq) - Ag(s)$ couple] that lies below it in the table of electrode potentials, but it will not react to a significant extent with the oxidized form of a couple that lies above it in the table.

Exercise 6.7. By (a) using the rule just stated above, and (b) by calculating the standard cell potential, determine whether, under standard conditions, $Fe^{3+}(aq)$ will oxidize $Cl^-(aq)$ to $Cl_2(g)$ to a significant extent.

Solution. (a) $Cl^-(aq)$ is the reduced form of $Cl_2(g)$. Therefore, it will react to a significant extent with the oxidized form of any couple that lies *below* it in Table 6.2. $Fe^{3+}(aq)$ is the oxidized (rust) form of $Fe(s)$, but it lies *above* $Cl^-(aq)$ in Table 6.2. Therefore, $Cl^-(aq)$ will not react to a significant extent with $Fe^{3+}(aq)$.[8] (b) The half-reactions and their electric potentials are, from Table 6.2

$$2Cl^-(aq) \rightarrow Cl_2(g) + 2e^- \qquad E^0_{ox} = -(1.36 \text{ V}) = -1.36 \text{ V}$$
$$2Fe^{3+}(aq) + 2e^- \rightarrow 2Fe^{2+}(aq) \qquad E^0_{red} = 0.771 \text{ V}$$

Net: $2Fe^{3+}(aq) + 2Cl^-(aq) \rightarrow \qquad E^0_{cell} = -0.589 \text{ V}$
$ Cl_2(g) + 2Fe^{2+}(aq)$

The negative value of E^0_{cell} confirms that the reaction will not proceed to a significant extent in the forward direction.

To balance the electrons in the oxidation and reduction half-reaction in Exercise 6.7, we had to multiply by 2 the reduction half-reaction given in Table 6.2. However, the magnitude of the electrode potential was unaffected. This is because the magnitude of E^0 is determined only by the concentrations of the species in the half-reaction (1 mole per liter under standard conditions), not by the amounts of the species (in moles).

6.6 Standard cell potentials and free-energy change

We saw in Section 2.2 that if a chemical transformation is spontaneous, and pressure and temperature are constant, the Gibbs free energy (G) of the system will decrease. We have seen in the above section that a redox reaction will proceed spontaneously under standard conditions if its standard cell potential (E^0_{cell}) is positive. Therefore, there should be a quantitative relationship between ΔG^0 under standard conditions and E^0_{cell}. The reader was asked to derive this relationship in Exercise 2.29 (the solution is given in Appendix VII); applied to standard states the relationship is

$$\Delta G^0 = -nFE^0_{cell} \qquad (6.20)$$

where F is the Faraday constant, which was defined in Exercise 6.5 and is equal to 96,489 coulombs. The value of n to be used in Eq. (6.20) for any particular redox reaction is determined by expressing the reaction as the sum of two half-reactions, in which electrons appear in equal numbers. The number of electrons in each half-reaction is then equal to the value of n. For example, for the redox reaction (6.18), $n = 2$, because there are two electrons in Eqs. (6.16) and (6.17).

Exercise 6.8. Using values of E^0 given in Table 6.2, determine the change in the Gibbs free energy under standard conditions for the redox reaction

$$2Cr(s) + 6H^+(aq) \rightarrow 2Cr^{3+}(aq) + 3H_2(g)$$

Is this a spontaneous reaction?
Solution. The half-reactions are

$2Cr(s) \rightarrow 2Cr^{3+}(aq) + 6e^-$	$E^0_{ox} = -(-0.74 \text{ V}) = 0.74 \text{ V}$
$6H^+(aq) + 6e^- \rightarrow 3H_2(g)$	$E^0_{red} = 0 \text{ V}$

Net: $2Cr(s) + 6H^+(aq) \rightarrow$ $\qquad E^0_{cell} = 0.74 \text{ V}$
$\qquad 2Cr^{3+}(aq) + 3H_2(g)$

Note that the oxidation half-reaction involving the $Cr(s) - Cr^{3+}(aq)$ couple given in Table 6.2 has been multiplied by 2, and the reduction half-reaction involving the $H^+(aq) - H_2(g)$ couple given in Table 6.2 has been multiplied by 3, in order to make the number of electrons released by the oxidation half-reaction equal to those consumed by the reduction half-reaction.

Since six electrons are involved in each of the balanced half-reactions, $n = 6$. Also, $E^0_{cell} = 0.74$ V and $F = 96,489$ coulombs. Therefore, from Eq. (6.20)

$$\Delta G^0 = -n FE^0_{cell} = -6(96,489)(0.74) = -4.3 \times 10^5 \text{ J}$$

Since ΔG^0 is negative (and E^0_{cell} is positive), the redox reaction is spontaneous.

In the above exercise we derived ΔG^0 from E^0_{cell}. Conversely, we can use Eq. (6.20) to derive the value of E_{cell}^0 for a reaction from the ΔG^o value, where ΔG^o for molar quantities is given by Eq. (2.34). We can also relate the standard cell potential for a reaction that is in chemical equilibrium to the equilibrium constant K_c for the reaction. Combining Eqs. (2.46) and (6.20), we get

$$-n F E_{cell}^0 = -R^* T \ln K_c$$

or,

$$E_{cell}^0 = \frac{2.30 \, R^* T}{nF} \log K_c \qquad (6.21)$$

or, substituting $R^* = 8.314$ J deg^{-1} mol^{-1} and $T = 298$K,

$$E_{cell}^0 = \frac{0.0591}{n} \log K_c \qquad (6.22)$$

where, as usual, E_{cell}^0 is in volts.

6.7 The Nernst equation

So far we have considered only standard cell potentials, that is, the electric potential difference developed by a chemical reaction that is at equilibrium in an electrochemical cell at normal atmospheric pressure and a temperature of 25°C, and when the chemical species are present in standard concentrations. We can derive an expression for the electric potential difference generated under nonequilibrium and nonstandard conditions (E_{cell}) as follows. If we write Eq. (2.41) in terms of concentrations and remove the requirement of molar concentrations, we get

$$\Delta G = \Delta G^0 + R^* T \ln \frac{[G]^g [H]^h \ldots}{[A]^a [B]^b \ldots} \qquad (6.23)$$

where [A], [B] . . . and [G], [H] are the concentrations of the reactants and products in the general chemical reaction given by Eq. (1.3), which may or may not be in equilibrium. Combining Eqs. (6.23) and (1.10)

$$\Delta G - \Delta G^0 + R^* T \ln Q \qquad (6.24)$$

where Q is the reaction quotient. From Eq. (6.20), $\Delta G = -n F E_{cell}$ and $\Delta G^0 = -n F E_{cell}^0$. Therefore, Eq. (6.24) becomes

$$E_{cell} = E_{cell}^0 - \frac{R^*T}{nF} \ln Q$$

or,

$$E_{cell} = E_{cell}^0 - \frac{2.30R^*T}{nF} \log Q \qquad (6.25)$$

This last relationship is called the *Nernst equation*. Substituting the values of R^* and F into Eq. (6.25), and taking $T = 298K$,

$$E_{cell} = E_{cell}^0 - \frac{0.0591}{n} \log Q \qquad (6.26)$$

Exercise 6.9. Calculate the initial electric potential difference generated in an electrochemical cell at 298K by the redox reaction

$$2Cr(s) + 6H^+(aq) \rightarrow 2Cr^{3+}(aq) + 3H_2(g)$$

if $[Cr(s)] = 0.5$ M, $[H^+(aq)] = 1$ M, $[Cr^{3+}(aq)] = 2$ M, and $[H_2(g)] = 1$ M.

Solution. In Exercise 6.8 we calculated that the electric potential difference generated by this reaction under standard conditions (i.e., E_{cell}^0) was 0.74 V; we also saw that $n = 6$. We can use the Nernst equation to derive the electric potential difference generated by the reaction under the nonstandard concentrations specified in the present exercise if we know the value of Q. Applying Eq. (1.10) to the reaction (remembering that the concentrations of liquids and solids are equated to unity in this expression), we get

$$Q = \frac{[Cr^{3+}(aq)]^2[H_2(g)]^3}{[H^+(aq)]^6} = \frac{(2)^2(1)^3}{(1)^6} = 4$$

Therefore, from Eq. (6.26)

$$E_{cell} = 0.74 - \frac{0.0591}{6} \log 4$$

or,

$$E_{cell}^0 = 0.74 - 0.006 = 0.73 \text{ V}$$

We saw in Exercise 6.8 that when all the species in the specified reaction were present in concentrations of 1 M, the cell potential was 0.74 V and the reaction was spontaneous from left to right. Exercise

6.9 shows that if the concentration of one of the reactants is decreased {[Cr(s)] went from 1 M to 0.5 M}, and the concentration of one of the products is increased {[Cr^{3+}(aq)] went from 1 M to 2 M}, the cell potential is decreased (from 0.74 V to 0.73 V in this case). This is because an increase in the concentration of one of the products and a decrease in the concentration of one of the reactants reduces the forward reaction. Left to itself, a chemical system will move toward its equilibrium state. When it reaches this state $K_c = Q$ (see Section 1.3) and, from Eqs. (6.22) and (6.26), $E_{cell} = 0$. We have also seen in Section 2.4 that when a chemical system is at equilibrium $\Delta G = 0$. Hence, when the species in an electrochemical cell are in equilibrium, the Gibbs free energy is a minimum and the cell generates zero electric potential difference (i.e., "the battery has run down").

6.8 Redox potentials; Eh–pH diagrams

In geochemistry, the ability of an environmental system to oxidize or reduce, as measured by its electrode potential, is often called its *redox potential*. It is given the symbol Eh (where "h" indicates that the reference half-cell is hydrogen). The Eh is analogous to pH, in that it is a measure of the ability of a system to supply or to take up electrons, while the pH of a system measures its ability to supply or take up protons. If a system has a propensity for *supplying* electrons (i.e., if it provides a good reducing environment for an oxidant), it will have a low value of Eh (i.e., be high up the list in Table 6.2). If a system has a propensity for *consuming* electrons (i.e., if it provides a good oxidizing environment for a reductant), it will have a high value of Eh (i.e., be low down in the list in Table 6.2).

There are limits to the ranges of values of Eh and pH in natural environments. For example, some of the most acidic solutions in nature, with pH values below zero, occur near active volcanoes due to the emissions of acidic gases. However, these acidities are quickly reduced by reactions with soils and rocks. Complete neutralization is generally not obtained because of the buffering capacity of CO_2 from the air (see Section 5.13). Thus, the pH values of natural systems generally lie in the range 5 to 6, with 4 as a rough lower limit. Very basic solutions, with pH values up to about 11, can form if CO_2-free water is in contact with carbonate rocks or certain silicates. Again, however, because CO_2 acts as a buffer, the pH values of surface waters generally do not rise above about 9.

We see from Table 6.2 that the reaction

$$2H_2O(l) \rightarrow O_2(g) + 4H^+(aq) + 4e^- \qquad (6.27)$$

has a strong oxidizing potential ($E^0_{ox} = -1.229$ V). Consequently, any system with a redox potential greater than 1.229 V can be reduced by water, with the liberation of oxygen. Thus, oxygen in air is the most common oxidant in natural environments. This is illustrated by the following example. Under standard conditions the $F_2(s) - F^-(aq)$ system has a redox potential of 2.87 V (see Table 6.2). Therefore, it is reduced by water

$$2H_2O(l) \rightarrow O_2(g) + 4H^+(aq) + 4e^- \qquad E^0_{ox} = -(1.229 \text{ V})$$
$$2F_2(s) + 4e^- \rightarrow 4F^-(aq) \qquad E^0_{red} = 2.87 \text{ V}$$

Net: $2H_2O(l) + 2F_2(s) \rightarrow O_2 + 4H^+(aq) + 4F^-(aq) \qquad E^0_{cell} = 1.641V$

where the positive value of E^0_{cell} indicates that the reaction is spontaneous from left to right.

For nonstandard conditions, the redox potential of the $O_2(g) - H_2O(l)$ system is, from Eq. (6.26)

$$E_{cell} = 1.229 - \frac{0.0591}{4} \log\frac{1}{[O_2(g)][H^+(aq)]^4}$$

$$= 1.229 + 0.0148 \log[O_2(g)] + 0.0591 \log[H^+(aq)]$$

Therefore, since in air $[O_2(g)] = 0.2$ atm and $\log[H^+(aq)] = -pH$

$$E_{cell} = 1.229 - 0.010 - 0.0591 \text{ pH}$$

or,

$$E_{cell} = 1.22 - 0.0591 \text{ pH} \qquad (6.28)$$

Thus, redox potentials in natural environments will generally be less that the value of E_{cell} given by Eq. (6.28).

In the limit considered above, water acted as a reductant to limit the oxidation that can occur in nature. Water also acts as an oxidant to limit the reduction that can occur in nature. In this case, the half-reaction involving water is

$$2H_2O(l) + 2e^- \rightarrow H_2(g) + 2OH^-(aq) \qquad (6.29)$$

which, from Table 6.2, has a standard electrode potential of -0.828 V. Under standard conditions, Reaction (6.29) can combine spontaneously with all the oxidation half-reactions in Table 6.2 that lie *above* it (since, when these pairs of half-reactions are combined, $E^0_{cell} > 0$). However, since water must be in equilibrium with $H^+(aq)$, and $OH^-(aq)$, and $OH^-(aq)$ are released by the half-reaction (6.29), the concentration of $H^+(aq)$ must decrease. This is achieved through the half-reaction

$$2H^+(aq) + 2e^- \rightarrow H_2(g) \tag{6.30}$$

which can combine spontaneously with any of the reverse half-reactions lying above it in Table 6.2 (since combinations of these half-reactions give $E^0_{cell} > 0$). Thus, the electrode potential corresponding to the half-reaction (6.30) defines the lower limit of Eh for natural systems. Using Eq. (6.26), this lower limit is

$$E_{cell} = E^0_{cell} - \frac{0.0591}{n} \log Q$$

$$= 0 - \frac{0.0591}{2} \log\frac{[H_2(g)]}{[H^+(aq)]^2}$$

$$= 0.0591 \log[H^+(aq)] - 0.0295 \log[H_2(g)]$$

Since, $pH = -\log[H^+(aq)]$, and the partial pressure of hydrogen gas near the Earth's surface cannot exceed 1 atm, the minimum value of E_{cell} for a system in the presence of water is

$$E_{cell} = -0.0591 \; pH \tag{6.31}$$

Thus, the redox potential in natural environments near the Earth's surface should not fall below the value given by Eq. (6.31).

The natural limits to pH and Eh discussed above are shown in Figure 6.2. When the pH and Eh values of other oxidation processes are plotted on this diagram, it can easily be seen over what ranges they can be expected to occur in natural environments.

6.9 Gram-equivalent weight and normality

Instead of moles for mass and molarity for concentration, *gram-equivalent weight* (or *equiv.*) and *normality* are sometimes used in considering redox reactions. The equiv. is the amount of a substance associated with 1 mole of electrons in a redox half-reaction. For example, in the half-reaction (6.16), 1 mole of $Zn(s)$ and 1 mole of $Zn^{2+}(aq)$ are associated with 2 moles of electrons. Therefore, 1/2 mole of $Zn(s)$

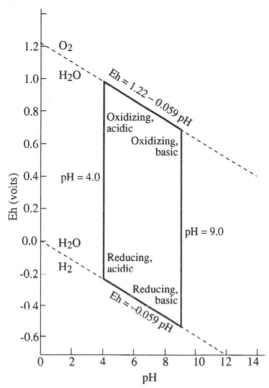

Figure 6.2. An Eh–pH diagram. For natural systems near the Earth's surface, the usual ranges of values of pH and Eh lie within the parallelogram. (From *Introduction to Geochemistry* by K. B. Krauskopf, McGraw Hill, 1967. Reproduced with permission of McGraw-Hill.)

is 1 equiv. of $Zn(s)$, and 1/2 mole of $Zn^{2+}(aq)$ is 1 equiv. of $Zn^{2+}(aq)$. In the half-reaction (6.17), 2 moles of $H^+(aq)$ and 1 mole of $H_2(g)$ are associated with 2 moles of electrons. Therefore, 1 mole of $H^+(aq)$ is 1 equiv. of $H^+(aq)$, and 1/2 mole of $H_2(g)$ is 1 equiv. of $H_2(g)$. Since the electric charge associated with the passage of 1 mole of electrons is the Faraday constant F (96,489 coulombs), *when a charge of 1 F passes through an electrochemical cell, it liberates or consumes one equiv. of each of the species involved in the chemical reaction.*

The *normality* (N) of a solution is the number of gram-equivalent weights of solute in 1 L of the solution.

Exercise 6.10. A certain quantity of electricity liberates 0.72 g of silver in an electrochemical cell containing a silver solution. If this same quantity of electricity deposits 0.44 g of gold when it passes through a gold solution, what is the oxidation state of the silver in the silver solution and the gold in the gold solution? (One equiv. of silver $= 108$ g.)

Solution. The number of equivalents of silver liberated by the passage of the electric charge is $0.72/108 = 0.0067$. Since this same electric charge will release 0.0067 equiv. of gold, 0.44 g of gold is the same as 0.0067 equiv. of gold. Therefore, 1 gram-equivalent weight of gold $= 0.44/0.0067 = 66$. Let the oxidation state of the silver in solution be n, then

$$Ag^{n+}(aq) + ne^- \rightarrow Ag(s)$$

Therefore,

$$1 \text{ equiv. of silver} = \frac{1 \text{ mole of silver}}{n}$$

or,

$$n = \frac{1 \text{ mole of silver}}{1 \text{ equiv. of silver}} = \frac{108}{108} = 1$$

Similarly, if the oxidation state of the gold is m

$$m = \frac{1 \text{ mole of gold}}{1 \text{ equiv. of gold}} = \frac{197}{66} = 3$$

(*Note:* A given oxidant or reductant may have more than one equivalence, depending on the reaction involved.)

Exercises

6.11.   Answer, interpret, or explain the following in light of the principles presented in this chapter.

(a) Hydrogen peroxide, $H_2O_2(g)$, is generally used commercially in the form of a 30% aqueous solution. The bottle contains the warning: "Contact with other materials may cause fire. If swallowed, give water or milk to drink."

(b) The reaction $Zn(s) + Cu^{2+}(aq) \rightarrow Zn^{2+}(aq) + Cu(s)$ is an oxidation–reduction reaction, even though oxygen is not involved.

(c) Sulfur dioxide gas, $SO_2(g)$, is emitted into the air when sulfur-containing coal or gas is burned in electric power plants. The $SO_2(g)$ dissolves in cloud drops to form a sulfurous acid, $H_2SO_3(aq)$, solution that is a strong reductant. The sulfurous acid solution is oxidized to a sulfuric acid, $H_2SO_4(aq)$, solution. Which gases, absorbed into cloud drops from the atmosphere, do you think may be involved in this oxidation?

(d) During the day the concentration of CO_2 in forests is about 305 ppm, but during the night it is about 340 ppm.

(e) The energy provided by burning fossil fuels (coal, oil, wood, etc.) derives from the sun.

(f) The burning of fossil fuels generally increases the CO_2 content of the atmosphere but the burning of recently grown trees does not.

(g) The surface of the Earth is a redox boundary between the planet's reduced metallic core and an oxidizing atmosphere.

(h) The tendency of most materials to be oxidized by air is counteracted by photosynthesis.

(i) Much chemical technology depends on the reduction of materials to lower oxidation states (e.g., H and NH_3), which are then reoxidized when used.

(j) If the values of x electrode potentials are known, it is possible to determine the spontaneous directions of $x(x-1)/2$ chemical reactions under standard conditions.

(k) A chemical reaction in an aqueous solution that has a negative value of E^0_{cell} can be driven in the forward direction by applying an electric potential difference that exceeds $|E^0_{cell}|$ in an appropriate direction. (This is the basis of electrolysis.)

(l) A voltmeter attached to the circuit of an electrochemical cell serves as a "Gibbs free-energy meter."

(m) Stronger oxidants than oxygen do not persist for long in natural environments.

(n) Equal numbers of gram-equivalents of two substances react exactly with each other.

6.12. Write down the half-reactions for the following oxidation–reduction reactions. In each case, indicate which is the oxidation half-reaction and which the reduction half-reaction, and which species is the oxidant and which the reductant.

(a) $Fe^{3+}(aq) + Cu^+(aq) \rightarrow Fe^{2+}(aq) + Cu^{2+}(aq)$

(b) $Zn(s) + 2H^+(aq) \rightarrow Zn^{2+}(aq) + H_2(g)$

6.13.   What are the oxidation numbers of the elements in the following compounds (a) $NO_2^-(aq)$, (b) $NO_3^-(aq)$ (c) $HNO_3(aq)$ and (d) $H_2SO_3(aq)$?

6.14.   Are the following redox reactions?
(a) $2SO_2(g) + O_2(g) \rightarrow 2SO_3(g)$
(b) $2CCl_4(aq) + K_2CrO_4(aq) \rightarrow$
$\qquad 2Cl_2CO(aq) + CrO_2Cl_2(aq) + 2KCl(aq)$

6.15.   Oxygen in water can oxidize lead. Balance the following redox reaction that represents this process:

$$Pb(s) + O_2(aq) + H_2O(l) \rightarrow Pb^{2+}(aq) + OH^-(aq)$$

6.16.   Balance the equation for the following redox reaction in an aqueous solution:

$$P_4(s) + OH^-(aq) + H_2O(l) \rightarrow H_2PO_2^-(aq) + PH_3(g)$$

6.17.   The unbalanced equation for the chemical reaction representing respiration is

$$C_6H_{12}O_6(s) + O_2(g) \rightarrow CO_2(g) + H_2O(l)$$

Which is the oxidant and which the reductant in this reaction?

6.18.   Arrange the following in order of increasing strengths as oxidants: $Ag^+(aq)$, $Mn^{2+}(aq)$, $O_3(aq)$, and $Fe^{2+}(aq)$.

6.19.   Use Table 6.2 to determine if, under standard conditions, (a) $Fe^{3+}(aq)$ will oxidize $Ni(s)$ to $Ni^{2+}(aq)$ to any significant extent, (b) $H_2O_2(aq)$ will behave as an oxidant or reductant with respect to the $SO_4^{2-}(aq) - S_2O_8^{2-}(aq)$ couple, (c) $H_2O(aq)$ will behave as an oxidant or reductant with respect to the $Ag^+(aq) - Ag(s)$ couple, and (d) $SO_4^{2-}(aq)$ will oxidize $Fe^{2+}(aq)$ to $Fe^{3+}(aq)$ to a significant extent?

6.20.   Determine whether the following reactions proceed spontaneously from left to right in acidic solutions under standard conditions. What are the values of their standard cell potentials?
(a) $H_2S(aq) + Cl_2(aq) \rightarrow S(s) + 2Cl^-(aq) + 2H^+(aq)$
(b) $4SO_4^{2-}(aq) + O_2(g) + 4H^+(aq) \rightarrow 2S_2O_8^{2-}(aq) + 2H_2O(l)$

6.21.   Using the appropriate values of E^0 given in Table 6.2, determine the initial change in the Gibbs free energy under standard conditions for the redox reaction

$$H_2O_2(aq) + S(s) \rightarrow O_2(g) + H_2S(aq)$$

Is this a spontaneous reaction?

6.22. What is the value of the equilibrium constant for the redox reaction given in Exercise 6.21? What do you conclude from this value?

6.23. Calculate the initial electric potential difference in an electrochemical cell at 250K, and the initial change in the Gibbs free energy (ΔG) associated with the redox reaction

$$Cd(s) + Cu^{2+}(aq) \rightarrow Cd^{2+}(aq) + Cu(s)$$

when the species are present in the following concentrations $[Cd(s)] = 3.00$ M, $[Cu^{2+}(aq)] = 1.50$ M, $[Cd^{2+}(aq)] = 0.500$ M, and $[Cu(s)] = 0.750$ M. Is the reaction spontaneous?

6.24. Automobile batteries often consist of six cells connected in series. The following redox reaction occurs in each cell

$$Pb(s) + PbO_2(s) + 4H^+(aq) + 2SO_4^{2-}(aq) \rightarrow$$
$$2PbSO_4(s) + 2H_2O(l)$$

What is the maximum voltage that this battery can generate under standard conditions?

6.25. The cell potential when a $Mn(s) - Mn^{2+}(aq)$ couple is linked to an $H^+(aq) - H_2(g)$ couple is 1.01 V at 25°C when both the $Mn(s)$ and the $Mn^{2+}(aq)$ are present in concentrations of 1.0 M and the partial pressure of the $H_2(g)$ is 1.0 atm. What is the concentration of $H^+(aq)$? [*Note:* This problem indicates how an electrochemical cell that involves $H^+(aq)$ can be used to measure $[H^+(aq)]$, and therefore the pH of a solution. This is the basis of a pH meter.]

6.26. Calculate the standard electrode potential for the half-reaction

$$HSO_3^-(aq) + H_2O(l) \rightarrow HSO_4^-(aq) + 2H^+(aq) + 2e^-$$

Is this reaction spontaneous under standard conditions?

6.27. What is the ratio of $[Fe^{2+}(aq)]$ to $[Fe^{3+}(aq)]$ that will be in equilibrium at 25°C with seawater that has a redox potential of 0.600 V?

6.28. $H_2S(aq) + HNO_3(aq) \rightarrow S(s) + NO(g) + H_2O(l)$ is an unbalanced redox reaction. What is the gram-equivalent weight of $H_2S(aq)$ and $HNO_3(aq)$?

6.29. What is the normality of 1.0 L of a solution that contains 35.6 g of HNO_3 if the only chemical reaction is ionization?

Notes

1 The reader might wonder why a reaction such as (6.3), which does not involve oxygen, is called an oxidation–reduction reaction. The reason is as follows. Originally the term "oxidation" was applied to reactions in which a substance combines with oxygen, for example, the oxidation of copper in air

$$2Cu(s) + O_2(g) \rightarrow 2CuO(s)$$

This reaction involves each copper atom losing two electrons, and each oxygen atom gaining two electrons

$$2Cu(s) \rightarrow 2Cu^{2+}(aq) + 4e^-$$
$$O2(g) + 4e^- \rightarrow 2O^{2-}(aq)$$

Net: $2Cu(s) + O_2(g) \rightarrow 2Cu^{2+}(aq) + 2O^{2-}(aq) \rightarrow 2CuO(s)$

Thus, by analogy, all chemical reactions involving simultaneous electron loss and electron gain came to be called oxidation–reduction reactions.

2 From the viewpoint of atomic structure, every atom consists of a positive nucleus surrounded by negative electrons. In forming chemical bonds .toms donate, receive, or share electrons. The number of electrons of an atom that is involved in this way in forming bonds with other atoms is the oxidation number of the atom. A positive oxidation number indicates that the atom has donated one or more electrons (e.g., $+1$ for the hydrogen atom in a molecule), and a negative oxidation number indicates that the atom has received one or more electrons (e.g., -2 for the oxygen atom in a molecule).

3 In redox reactions it is better to indicate the aqueous proton by $H^+(aq)$ rather than by $H_3O^+(aq)$, since it decreases the number of water molecules that must be written.

4 For example, in the half-cell on the right side of Figure 6.1, $Cu^{2+}(aq)$ ions are released from the copper electrode into the solution. These attract $SO_4^{2-}(aq)$ ions with which they combine: $Cu^{2+}(aq) + SO_4^{2-}(aq) \rightarrow CuSO_4(aq)$. This would leave the solution with a net positive charge if it were not for negative ions drifting into this half-cell from the salt bridge. In the half-cell on the left side of Figure 6.1, $Ag^+(aq)$ ions from the solution are deposited onto the silver electrode. This leaves $NO_3^-(aq)$ ions in the vicinity of this electrode; these attract $Ag^+(aq)$ ions with which they combine: $NO_3^-(aq) + Ag^+(aq) \rightarrow AgNO_3(aq)$. Again, the cell is maintained electrically neutral by positive ions drifting into it from the salt bridge.

5 A 1 M solution is in its standard state only if it behaves ideally (see Section 4.4), but this complication need not concern us here.

6 The number of electrons in 1 mole of electrons is Avogadro's number (6.0229×10^{23}), and the charge on one electron is 1.6021×10^{-19} coulombs. Therefore, the charge on 1 mole of electrons is $(6.0229 \times 10^{23}) \times (1.6021 \times 10^{-19})$ or 96,489 coulombs; this is the *Faraday constant* (F).

7 The following is an alternative explanation for the fact that type (a) combinations are significant (i.e., spontaneous) and type (b) combinations are not. In type (a) combinations the better oxidant serves as the oxidant, and the better reductant serves as the reductant. However, in type (b) combinations the better oxidant is forced into the role of a reductant and the better reductant into the role of the oxidant.

8 Of course, $Cl^-(aq)$ will react with $Fe^{3+}(aq)$ until a small amount of $Cl_2(g)$ is formed to bring the reaction to equilibrium. Therefore, the phrase "$Cl^-(aq)$ will not react to a significant extent with $Fe^{3+}(aq)$" really means "the ratio of $[Cl(aq)]$ to $[Fe^{2+}(aq)]$ is very large at equilibrium."

7

Photochemistry

A molecule may absorb electromagnetic (em) radiation and, in the process, break down into its atomic or molecular components. Unstable atoms and molecular fragments may also combine to form more stable molecules, disposing of their excess energy in the form of em radiation. These chemical reactions are called *photochemical*, and the process by which a photochemical reaction occurs is called *photolysis*. Photochemical reactions play very important roles in many aspects of environmental chemistry. Therefore, this book concludes with a brief account of some of the basic principles of photochemistry, which we will then apply to ozone in the Earth's stratosphere and the problem of the stratospheric ozone hole.

7.1 Some properties of electromagnetic waves

Electromagnetic radiation has both wave and particle characteristics. Considered as a wave, em radiation may be viewed as an ensemble of waves that travels through a vacuum with the speed of light, $c = 2.998 \times 10^8$ m s^{-1}. The distance between two successive crests in the intensity of the radiation is called the *wavelength* (λ) of the radiation. The *frequency* (ν) of the radiation is the number of crests in intensity that pass a given point in one second. The units of ν are s^{-1} or hertz (Hz). If ν crests in intensity pass a given point in one second, one crest passes a given point in $1/\nu$ seconds. Therefore, a crest in intensity of an em wave travels a distance λ in $1/\nu$ seconds. Hence, since speed = distance ÷ time,

$$c = \nu \, \lambda \tag{7.1}$$

The em spectrum is shown in Figure 7.1. As can be seen, a portion of the spectrum constitutes *visible light*. Within this portion, different

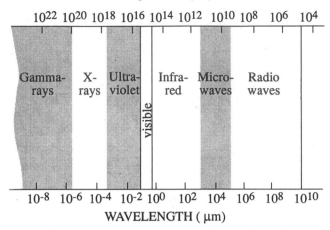

Figure 7.1. The electromagnetic spectrum. (Scales are logarithmic.)

wavelengths of em radiation produce different colors (e.g., red light is em radiation with wavelengths between about 0.620 and 0.760 μm, and dark blue light has wavelengths between 0.455 and 0.485 μm). Other parts of the spectrum are referred to as infrared (IR), ultraviolet (UV), X-ray radiation, etc.

Exercise 7.1. How many wavelengths of green light, with a frequency of 5.7×10^{14} s^{-1}, are there in 1 m?

Solution. From Eq. (7.1), the wavelength of the green light is

$$\frac{c}{\nu} = \frac{2.998 \times 10^8}{5.7 \times 10^{14}} = 0.53 \times 10^{-6} \text{m (or } 0.53\mu\text{m)}$$

Therefore, the number of wavelengths of green light in 1 m is

$$\frac{1}{0.53 \times 10^{-6}} = 1.9 \times 10^6$$

(*Note:* The number of wavelengths of em radiation in 1 m is called the *wave number* $\bar{\nu}$; the wave number is equal to $1/\lambda$.)

From the viewpoint of its particle character, the energy of an em wave is absorbed, emitted, or converted into other forms of energy in discrete units (or quanta) called *photons*. The amount of energy (W) associated with one photon of radiation is

$$W = h\nu \qquad (7.2)$$

where h is the Planck constant $(6.6262 \times 10^{-34}$ J s). From Eqs. (7.1) and (7.2)

$$W = \frac{hc}{\lambda} = hc\bar{\nu} \tag{7.3}$$

Thus, the amount of energy in a photon of em radiation is inversely proportional to the wavelength and directly proportional to the wave number of the radiation.

In photochemistry we are often interested in an Avogadro's number $(N_A = 6.022 \times 10^{23})$ of photons, which could be considered as "1 mole of photons"; this is called an *einstein*. One einstein of photons has energy

$$E_e = N_A h\nu \tag{7.4}$$

7.2 Some photochemical terminology and principles

An important principle of photochemistry is that each photon absorbed by a molecule activates that molecule in the initial (or *absorption*) step of a photochemical process. For example, the absorption of a photon by a general molecule, indicated by XY, can be represented by

$$XY \mid h\nu \rightarrow XY^* \tag{7.5}$$

where $h\nu$ is used to indicate one photon [$h\nu$ is actually the energy of one photon – see Eq. (7.2)] and XY^* indicates that the molecule XY has been raised to an electronically *excited state*. If one mole of XY is considered in Reaction (7.5), that is N_A molecules of XY, then 1 einstein of photons (i.e., N_A photons) would be involved.

Exercise 7.2. In sunlight, NO_2 in the air undergoes dissociation

$$NO_2(g) + h\nu \rightarrow NO_2^*(g) \rightarrow NO(g) + O(g) \tag{7.6}$$

If the energy of dissociation for this reaction is 304 kJ mole^{-1}, what is the minimum wave number of the EM radiation that can cause this dissociation?

Solution. If we consider the dissociation of 1 mole of NO_2, we will need to have 1 einstein of photons, which will need to have an energy of at least 304 kJ to produce the dissociation. Therefore, from Eq. (7.4)

$$N_A h\nu \geq 304 \times 10^3$$

or,

$$\nu \geq \frac{304 \times 10^3}{(6.022 \times 10^{23})(6.6262 \times 10^{-34})}$$

$$\nu \geq 7.62 \times 10^{14} \text{ Hz}$$

Therefore, from Eq. (7.1)

$$\frac{c}{\lambda} \geq 7.62 \times 10^{14}$$

or,

$$\lambda \leq \frac{c}{7.62 \times 10^{14}} = \frac{2.998 \times 10^8}{7.62 \times 10^{14}} = 0.39 \times 10^{-6} \text{m or } 0.39 \ \mu\text{m}$$

Since wave number $= 1/\lambda$, the minimum required wave number of the radiation is 2.6×10^6 m^{-1} or 2.6 μm^{-1}. (Dissociation of some NO_2 may occur at somewhat smaller wave numbers than this due to internal energy already present in the NO_2 molecules.)

Reaction (7.6) provides an example of one pathway that an excited molecule may follow, that is, to dissociate chemically. This pathway can be indicated in general by

$$XY^* \rightarrow A + B \text{ (dissociation)} \tag{7.7}$$

where A and B represent the products formed from the breakup of XY^*, which may be chemically different from X and Y and which may or may not be excited. This type of dissociation can occur if the energy of the absorbed photon exceeds the *binding energy* of the chemical bond for XY. The binding energy of chemical bonds are sometimes expressed in electron-volts (eV), where 1 eV $= 1.6 \times 10^{-19}$ J.[1] If 1 mole of a substance is considered, the units of binding energy are eV mole^{-1}, where 1 eV mole$^{-1} = (1.6 \times 10^{-19})N_A = 9.6 \times 10^4$ J mole^{-1}.

There are other possible pathways that may be taken by the excited molecule XY^* in Reaction (7.7). These can be represented by

$XY^* + CD \rightarrow E + F + - -$	(reaction)	(7.8)
$XY^* \rightarrow XY^+ + e^-$	(photoionization)	(7.9)
$XY^* \rightarrow XY + h\nu$	(luminescence)	(7.10)
$XY + XY^* \rightarrow XY + XY^\S$	(intramolecular energy transfer)	(7.11)
$XY^* + GH \rightarrow XY + GH^*$	(intermolecular energy transfer)	(7.12)
$XY^* + M \rightarrow XY + M$	(quenching)	(7.13)

Reactions (7.7) to (7.13) are referred to as *primary photochemical steps*. Reaction (7.8) represents a chemical reaction between XY* and a molecule CD leading to products E, F, – –, any of which can be in an excited state. In Reaction (7.9), the excited electron in XY* escapes to become a free electron leaving XY ionized; this is called *photoionization*. The electrons in the Earth's upper atmosphere (which permit long-distance radio communications) are produced primarily by solar radiation photoionizing molecules and atoms in the air. In Reaction (7.10), the radiation absorbed in the initial step (7.5) is re-emitted; this is called *luminescence*. If the radiation is re-emitted very quickly it is called *fluorescence;* if it is re-emitted slowly ($\sim 10^{-3}$ to 10^{-2} s) it is called *phosphorescence*. Fluorescence is primarily responsible for the phenomenon of *airglow*, that is, the emission of a faint glow from the Earth's upper atmosphere. In Reaction (7.11), the collision of XY* with another XY molecule transforms XY* to a new excited state (indicated by XY§). In Reaction (7.12), XY* interacts with another molecule GH which it excites. In Reaction (7.13), XY* interacts with a chemically unreactive molecule (M) to which it transfers its excited energy and is itself reduced to the electronic ground state.

In addition to the molecules that are specifically indicated as being in an excited state in Reactions (7.7) to (7.13), the other molecules and atoms in these reactions may or may not be excited. The energy change associated with the initial step in a photochemical process [Reaction (7.5)] generally involves the excitation of one electron in the molecule (or atom) from a lower to a higher energy level, although longer wavelengths may excite molecular vibrations or rotations. Transitions in the electronic, vibrational, and rotational energy states of a molecule are governed by the rules of quantum mechanics, which determine the characteristic frequencies of em radiation that a molecule can absorb or emit (e.g., the characteristic yellow light emitted by sodium compounds when they are heated in a flame is produced by one of these transitions). The excited state of an atom or a molecule can be indicated by spectroscopic notation, which is often placed in brackets following the chemical symbol. For example, O(^{3}P) indicates the ground state of the oxygen atom and O(^{1}D) an excited state. However, since it is beyond the scope of this book to explain the exact meaning of this notation, we will not use it.

7.3 Quantum yields

Although each photon absorbed by a molecule activates that molecule in the *initial* step of a photochemical process, it can be seen from Reactions (7.5) and (7.7) to (7.13) that this does not necessarily result in a chemical change. For example, the energy may be reradiated [Reaction (7.10)] or it may be transferred to another molecule [Reaction (7.13)]. This "inefficiency" factor in photochemistry is taken into account quantitatively by defining quantum yields.

The *overall quantum yield* of a stable product A is defined as

$$\Phi_A = \frac{\text{Number of moles of A formed}}{\text{Number of einsteins absorbed}} \qquad (7.14)$$

We can also define the quantum yield for a particular primary pathway (e.g., dissociation or reaction). Thus, the quantum yield for the ith primary pathway or process is

$$\phi_i = \frac{\text{Number of excited moles that proceed via pathway i}}{\text{Number of einsteins absorbed}} \qquad (7.15)$$

The probability of an excited molecule absorbing more than one photon during its short lifetime is very small for low to moderate light intensities (including all those in the atmosphere). Therefore,

$$\Sigma \phi_i = 1 \qquad (7.16)$$

where the summation includes all photochemical and photophysical primary pathways. Equation (7.16) is known as the *Stark–Einstein law*. Although $\phi_i \leq 1$, the overall quantum yield for a product (Φ_A) can be very much greater than unity, as illustrated in the following exercise.

Exercise 7.3. One of the very early photochemical reactions to be recognized was that involving hydrogen, $H_2(g)$, and chlorine, $Cl_2(g)$. Hydrogen and chlorine can be mixed together without reacting; but if the mixture is exposed to light the two substances react explosively to form hydrochloric acid, $HCl(g)$. The reaction sequence is

$$Cl_2(g) + h\nu \rightarrow 2Cl(g) \qquad (7.17a)$$
$$Cl(g) + H_2(g) \rightarrow HCl(g) + H(g) \qquad (7.17b)$$
$$H(g) + Cl_2(g) \rightarrow HCl(g) + Cl(g) \qquad (7.17c)$$

If Reactions (7.17b) and (7.17c) occur one million times for each Cl(g) molecule produced in Reaction (7.17a), what is the overall quantum yield of HCl?

Solution. Reactions (7.17) are an example of a *chemical chain reaction*. The chain is initiated by Reaction (7.17a); then Reactions (7.17b) and (7.17c) continue to produce HCl(g) until either the H(g) or Cl(g) is removed (by, for example, H + H → H_2 or Cl + Cl → Cl_2). However, if the Reactions (7.17b) and (7.17c) occur 10^6 times for each Cl(g) molecule produced by Reaction (7.17a), 10^6 moles of HCl(g) will be produced for every einstein of radiation absorbed in Reaction (7.17a). Therefore, $\Phi_{HCl} = 10^6$.

7.4 Rate coefficients for photolysis

Following the ideas discussed in Section 3.1, we could express the reaction rate for a photochemical reaction, such as Reaction (7.6), in the form

$$\text{Reaction rate} = k[NO_2(g)]^m[h\nu]^n \qquad (7.18)$$

where k is the rate coefficient and $m + n$ is the overall order of the reaction. Experiments show that for Reaction (7.6), $m = n = 1$. However, as we have seen, the reaction rate for a photochemical reaction generally varies sharply with the energy of the photon $h\nu$, and it will also depend on the flux of the photons and the quantum efficiency. This complication can be avoided if we consider a constant flux of photons with a fixed distribution with respect to wavelength and combine this with k to give

$$\text{Reaction rate} = j\,[NO_2(g)] \qquad (7.19)$$

where j is a pseudo first-order rate coefficient called the *photolytic rate coefficient*. (*Note:* Photolytic rate coefficients are generally represented by the symbol j, and are sometimes called *j-values*.)

Exercise 7.4. When the sun is overhead in midlatitudes, a typical value of j in Eq. (7.19) is 5.0×10^{-3} s^{-1}. If Reaction (7.6) is the only sink for $NO_2(g)$, what is its residence time in the atmosphere under these conditions? Under these same conditions, how many molecules of NO(g) per cubic meter of air at 1.0 atm and 20°C will be produced in 1 hour by Reaction (7.6) if the concentration of $NO_2(g)$ is 0.50 parts per billion by volume (ppbv) of air?

Solution. From Eq. (3.14) the residence time of NO_2 in air under the stated conditions would be $1/j = 1/(5.0 \times 10^{-3}) = 200$ s. From Eq. (7.19)

$$\text{Rate of production of } NO(g) = j\,[NO_2(g)]$$
$$= 5.0 \times 10^{-3}[NO_2(g)]$$

where the rate of production of $NO(g)$ is in molecules $m^{-3}\ s^{-1}$ if $[NO_2(g)]$ is in molecules m^{-3}. The total number of molecules in 1 m^3 of air (n_0) at pressure p (in pascals) and temperature T (in K) is given by Eq. (1.8g)

$$p = n_0\,k\,T$$

where k is the Boltzmann constant (1.381×10^{-23} J deg^{-1} molecule^{-1}). Therefore, at p = 1 atm = 1,013 mb = 1.013×10^5 Pa, and $T = 293$K,

$$n_0 = \frac{1.013 \times 10^5}{(1.381 \times 10^{-23})\,293}$$
$$= 2.5 \times 10^{25} \text{ molecule } m^{-3}$$

Since $NO_2(g)$ comprises 0.50 ppbv of air

$$[NO^2(g)] = (0.50 \times 10^{-9})(2.5 \times 10^{25})$$
$$= 1.3 \times 10^{16} \text{ molecule } m^{-3}$$

Therefore,

$$\text{Rate of production of } NO(g) = (5.0 \times 10^3)(1.3 \times 10^{16})$$
$$= 6.5 \times 10^{13} \text{ molecule } m^{-3}\ s^{-1}$$

Therefore, in 1 hour the number of molecules of $NO(g)$ produced in 1 m^3 by Reaction (7.6) is $(1 \times 60 \times 60)(6.5 \times 10^{13}) = 2.3 \times 10^{17.}$

7.5 Photostationary states

If a chemical system in equilibrium is placed in a beam of em radiation that is absorbed by one of the reactants, the rate of the forward reaction will be changed, thereby disturbing the equilibrium of the system. The system will therefore adjust itself until the forward and reverse rates are again in equilibrium. This is called the *photostationary state* of the system.

To illustrate the concept of a photostationary state, let us consider the simplest system for the equilibrium of $NO(g)$, $NO_2(g)$, and $O_3(g)$ in the lower part of the Earth's atmosphere (i.e., the troposphere). This equilibrium is governed by the following three reactions

$$NO_2(g) + h\nu \xrightarrow{j} NO(g) + O(g) \qquad (7.20)$$

$$O(g) + O_2(g) + M \xrightarrow{k_1} O_3(g) + M \qquad (7.21)$$

$$O_3(g) + NO(g) \xrightarrow{k_2} NO_2(g) + O_2(g) \qquad (7.22)$$

where the symbols above the arrows indicate the rate coefficients. Each of the above three reactions is rapid, and as NO(g) accumulates $NO_2(g)$ is formed by Reaction (7.22) as rapidly as it is depleted by the photolytic Reaction (7.20). Therefore,

$$j[NO_2(g)] = k_2[O_3(g)][NO(g)]$$

or,

$$\frac{[NO_2(g)]}{[NO(g)]} = \frac{k_2}{j}[O_3(g)] \qquad (7.23)$$

Equation (7.23) determines the ratio of the concentration of $NO_2(g)$ to NO(g) when this system is in a photostationary steady state.

It is worthwhile noting here that Reactions (7.20) and (7.21) are the only definitely established chemical mechanism for producing ozone in the troposphere; the other source for tropospheric ozone is downward transport from the stratosphere. Together these two sources maintain a background concentration of ozone in the troposphere of about 0.03 to 0.05 parts per million of the air by volume (ppmv). Ozone is of critical importance in the chemistry of the troposphere because, not only is it a powerful oxidant itself, it is the primary source of the *hydroxyl radical*[2] (OH), which is highly reactive and of paramount importance in tropospheric chemistry. Also, as can be seen from Eq. (7.23), the concentration of ozone in the air determines the ratio of $[NO_2(g)]$ to [NO(g)]. Nitric oxide, NO(g), is also a very reactive gas and of great importance in atmospheric chemistry.[3]

7.6 Stratospheric ozone and photochemistry; depletion of stratospheric ozone

In recent years a great deal of scientific attention and public concern has been directed at the problem of ozone depletion in the Earth's stratosphere caused by anthropogenic chemicals. Ozone is toxic, and it is a powerful air pollutant in the lower atmosphere. However, most of the ozone in the atmosphere is located in the stratosphere, which extends from about 10 to 15 km (depending on latitude) up to about 45 km above the Earth's surface. The presence of adequate concentrations of ozone in the stratosphere is essential for plant and animal life on Earth as we know it. This is because stratospheric ozone absorbs the most biologically harmful of the sun's UV radiation (called UV-B, which has a wavelength from 0.29 to 0.32 μm). Thus, decreases in stratospheric ozone are accompanied by increases in the intensity of UV-B radiation at the Earth's surface and (disproportionate) increases in biological cell damage,[4] which can lead to skin cancer and damage to plants. Since both the formation and depletion of ozone in the stratosphere involve photochemical reactions, we will conclude this chapter with a brief description of how ozone is formed in the stratosphere and some of the mechanisms by which it can be depleted both naturally and by anthropogenic emissions.

A simple chemical scheme for maintaining steady-state concentrations of ozone in an "oxygen-only" stratosphere was proposed by the geophysicist Sidney Chapman in 1930. The reaction scheme is[5]

$$O_2 + h\nu \xrightarrow{j_a} O + O \tag{7.24}$$

$$O + O_2 + M \xrightarrow{k_b} O_3 + M \tag{7.25}$$

$$O_3 + h\nu \xrightarrow{j_c} O_2 + O \tag{7.26}$$

$$O + O_3 \xrightarrow{k_d} O_2 + O_2 \tag{7.27}$$

Until the early 1960s it appeared that these reactions could explain the main features of the steady-state distribution of ozone in the stratosphere. However, subsequent and more refined measurements of the rate coefficients for Reactions (7.24) to (7.27) showed that the Chapman reactions generate ozone five times faster than they destroy it. This is due primarily to the low value of k_d in Reaction (7.27). Since the concentration of ozone in the stratosphere is not increasing at a rapid rate, there must be a much faster route for destroying ozone than indicated by the Chapman reactions. The search for this fast route, and the discovery of the sensitivity of stratospheric ozone concentrations to the presence of quite small amounts of certain trace chemicals, represents one of the most exciting and important areas of research in the environmental sciences over the past several decades. A brief account of current views on this subject is given below.[6]

Ozone is continually generated in the stratosphere by Reactions (7.24) and (7.25). Therefore, if the trace chemicals responsible for the removal of stratospheric ozone are not to be rapidly depleted, they must serve as a catalyst for the removal of ozone and/or atomic oxygen [note that the removal of atomic oxygen will slow down the production of ozone by Reaction (7.25)]. Most of the catalytic reactions that have been proposed for this purpose are of the form

$$X + O_3 \rightarrow XO + O_2 \qquad (7.28a)$$
$$XO + O \rightarrow X + O_2 \qquad (7.28b)$$

$$\text{Net:} \qquad O + O_3 \rightarrow O_2 + O_2 \qquad (7.29)$$

where X represents the catalyst. Reactions (7.28) form a cycle in which X is consumed in (7.28a) and reformed in (7.28b), with XO acting as an intermediate. Provided that both Reactions (7.28a) and (7.28b) are fast, Reaction (7.29) can proceed much faster than it would by the direct route proposed by Chapman [i.e., faster than Reaction (7.27)]. Under these conditions, and provided there is no appreciable sink for X, just a few molecules of X have the potential to eliminate indefinite numbers of ozone molecules and atomic oxygen.

In the natural (i.e., anthropogenically undisturbed) stratosphere, the most important contenders for the catalyst X are H, OH, NO, and Cl. For example, in the case of NO

$$NO + O_3 \rightarrow NO_2 + O_2 \qquad (7.30a)$$
$$NO_2 + O \rightarrow NO + O_2 \qquad (7.30b)$$

$$\text{Net:} \qquad O + O_3 \rightarrow O_2 + O_2 \qquad (7.31)$$

At a temperature of $-53°C$ (which is typical of the stratosphere), the rate coefficients for Reactions (7.30a) and (7.30b) are 3.5×10^{-15} and 9.3×10^{-12} cm^3 molecule^{-1} s^{-1}, respectively, compared to 6.8×10^{-16} cm^3 molecule^{-1} s^{-1} for k_d in Reaction (7.27). Hence, the rate constants for Reactions (7.30) are greater than those for the direct Reaction (7.27). However, whether or not Reactions (7.30) will destroy ozone faster than Reaction (7.27) will depend on the concentrations of NO_2 and O_3, as illustrated in the following exercise.

Exercise 7.5. If Reaction (7.30b) is the rate-determining step in the reaction cycle (7.30), derive an expression, in terms of rate coefficients and the concentration of ozone, for the concentrations that NO_2 must exceed if the reaction cycle (7.30) is to destroy ozone faster than Reaction (7.27).

Solution. Let k be the rate coefficient for Reaction (7.30b). Since this is the rate-determining step, the net Reaction (7.31) cannot proceed faster than the rate at which Reaction (7.30b) destroys atomic oxygen. This rate is given by

$$-\frac{d[O]}{dt} = k[NO_2][O] \tag{7.32}$$

The rate at which atomic oxygen is destroyed in Reaction (7.27) is

$$-\frac{d[O]}{dt} = k_d[O][O_3] \tag{7.33}$$

Therefore, if atomic oxygen, and therefore ozone, is to be destroyed faster by the reaction cycle (7.30) than by Reaction (7.27), the right side of Eq. (7.32) must exceed the right reaction side of Eq. (7.33), that is

$$k[NO_2][O] > k_d[O][O_3]$$

or

$$[NO_2] > \frac{k_d}{k}[O_3]$$

When the appropriate rate coefficients and concentrations for the various reactions and species in the stratosphere are taken into account, it appears that the catalytic cycles involving H, OH, NO, and Cl all make major contributions to the destruction of ozone in the stratosphere. Reaction cycle (7.30) dominates in the lower stratosphere; the cycles involving H and OH dominate in the upper stratosphere; and the cycle involving Cl is important in the middle stratosphere. However, the destruction of ozone by the various catalytic cycles is not simply an additive process. This is because the species in one cycle can react with those in another cycle. For example, two important interactions that affect the contributions of the OH and Cl cycles are

$$HO_2 + NO \rightarrow OH + NO_2$$

and,

$$ClO + NO \rightarrow Cl + NO_2$$

Other important competitive processes operate in the stratosphere, including "reservoir" species, which can divert potential catalysts and other compounds from active to inactive forms, but which "wait in the wings for their turn on the stage." Despite these complications, advanced chemical reaction schemes incorporated into numerical models can reproduce the vertical profiles of ozone and other chemical species in the "natural" stratosphere with some fidelity.

The natural concentrations of H, OH, NO, and Cl (most of which originate at the Earth's surface) in the stratosphere serve to catalyze Reaction (7.29) and to maintain approximately steady-state concentrations of ozone. However, if the concentrations of the catalysts X in Reaction (7.28) are increased significantly by anthropogenic activities, the delicate balance between the sources and sinks of atmospheric ozone will be disturbed, and stratospheric ozone concentrations can be expected to decrease. One of the first concerns in this respect were aircraft flying in the stratosphere (particularly supersonic aircraft). This is because aircraft engines emit nitric oxide (NO) which, as shown by Reactions (7.30), can serve as the X in Reaction (7.28). However, there are not sufficient numbers of aircraft flying in the stratosphere at the present time to perturb stratospheric ozone significantly.

Of much greater concern, with already documented impacts, is the catalytic action of chlorine, from chlorofluorocarbons, in depleting stratospheric ozone. Chlorofluorocarbons are compounds containing chlorine, fluorine, carbon, and sometimes hydrogen. They are commonly known as "Freons," of which Freon 11 ($CFCl_3$) and Freon 12 (CF_2Cl_2) are the most common. Freons were first synthesized in 1930, as the result of a search for a nontoxic, nonflammable refrigerant. Over the next half-century they became widely used, not only as refrigerants, but as propellants in aerosol cans, in plastic foam, and as solvents and cleansing agents. Concern about their effects on the atmosphere began in 1973 when it was found that Freons were spreading globally and, because of their inertness, were expected to have lifetimes of up to several hundred years in the troposphere.

Such long-lived compounds eventually find their way into the stratosphere. Here they absorb UV radiation in the wavelength interval 0.19–0.22 μm and photodissociate

$$CFCl_3 + h\nu \rightarrow CFCl_2 + Cl \qquad (7.34)$$

and,

$$CF_2Cl_2 + h\nu \rightarrow CF_2Cl + Cl$$

Therefore, it was argued, the chlorine atom freed by these reactions could serve as the catalyst X in Reactions (7.28) and destroy ozone in the cycle

$$Cl + O_3 \rightarrow ClO + O_2 \qquad (7.36a)$$
$$ClO + O \rightarrow Cl + O_2 \qquad (7.36b)$$

Net: $\qquad O_3 + O \rightarrow O_2 + O_2 \qquad (7.37)$

The first evidence for depletions in stratospheric ozone produced by anthropogenic chemicals in the stratosphere came, surprisingly, from measurements over the Antarctic. In 1985 British scientists who had been making ground-based, remote-sensing measurements of ozone at Halley Bay (76°S) in the Antarctic for many years reported that there had been about a 30% decrease in total column ozone each October (i.e., in the austral spring) since 1977. These observations were subsequently confirmed by remote-sensing measurements from satellite and by airborne measurements. Satellite measurements show that the region of depleted ozone over the Antarctic in spring has grown progressively deeper since 1979, and in 1987 through 1991 it occupied an area larger than the Antarctic continent.

Detection of the so-called "ozone hole" over the Antarctic raised several intriguing questions. Why over the Antarctic? Why during spring? Also, the magnitudes of the measured decreases in ozone over the Antarctic were much greater than any predictions based solely on gas-phase chemistry, of the type outlined above – why? The answers to these questions provide an excellent demonstration of the maxim that in the environment processes rarely, if ever, act in isolation.

During the austral winter (June–September) stratospheric air over the Antarctic continent is restricted from interacting with air from lower latitudes by a large-scale vortex circulation, which is bound at its perimeter by strongly circulating winds, through which very cold air slowly sinks. High-level clouds, called polar stratospheric clouds (PSCs), form in the cold core of this vortex, where temperatures can fall below –80°C. In the austral spring, as temperatures rise, the winds around the vortex weaken, and by November the vortex has disappeared. However, during winter the vortex serves as a giant chemical reactor in which anomalous chemistry can go on. For example, although the concentrations of ozone in the vortex are normal in August, the concentrations of ClO are ten times greater than just outside the "wall" of the vortex and, by September, ozone concentrations within the vortex have decreased dramatically. There are also sharp decreases in the oxides of nitrogen (NO_y) and in the water-vapor content of the air when passing from outside to inside the wall of the vortex. The denitrification and dehydration are due, respectively, to the conversion of NO_y to nitric acid and the condensation of water at the very low temperatures inside the vortex. These two condensates form two types of PSCs. One type consists of nitric acid trihydrate particles, about 1 μm in diameter, which condense at about –80°C. The other type of PSC consists of ice–water particles (with nitric acid dissolved in them), about 10 μm in diameter, which condense near – 85°C. As the particles in these clouds slowly sink, they remove both water and nitrogen compounds from the stratosphere. As we shall see below, these processes play important roles in depleting ozone concentrations in the Antarctic vortex.

Let us now return to the chemistry associated with the depletion of stratospheric ozone. Most of the chlorine and chlorine oxide released into the stratosphere by Reactions (7.34) and (7.35) are quickly tied up in reservoirs as hydrogen chloride and chloride nitrate by the reactions

$$Cl + CH_4 \rightarrow HCl + CH_3 \tag{7.38}$$

$$ClO + NO_2 + M \rightarrow ClONO_2 + M \tag{7.39}$$

Liberation of the active chlorine atoms from these reservoirs is generally slow. However, on the surface of the ice particles that form PSCs, the following catalytic (catalyzed by the ice) reaction can occur

$$ClONO_2(s) + HCl(s) \rightarrow Cl_2(g) + HNO_3(s) \tag{7.40}$$

where the parenthetical "s" has been inserted to emphasize those compounds that are on (or in) ice particles. The nitric acid remains in the ice particles, but Cl_2 is released as a gas that is photodissociated in the stratosphere

$$Cl_2 + h\nu \rightarrow Cl + Cl \tag{7.41}$$

In addition to catalyzing Reaction (7.40), the ice particles play another role: they remove nitrogen from the stratosphere (as HNO_3), which limits the forward Reaction of (7.39), thereby providing more ClO than would otherwise be available. Thus, on both counts, during the austral winter the ice particles that comprise PSCs in the Antarctic vortex set the stage for the destruction of ozone by enhancing the concentrations of active ClO and Cl. However, Reactions (7.36) cannot proceed with full vigor until enough sunlight appears to release both sufficient free chlorine atoms [by Reaction (7.41)] *and* sufficient quantities of atomic oxygen [by Reaction (7.26)]. Since the latter requirement is not met in early spring in the Antarctic stratosphere, Reactions (7.36) cannot explain the very large depletions of ozone that produce the Antarctic ozone hole, although they probably contribute to it.

A cycle, catalyzed by ClO, that does appear capable of explaining the Antarctic ozone hole is

$$ClO + ClO + M \rightarrow (ClO)_2 + M \qquad (7.42a)$$
$$(ClO)_2 + h\nu \rightarrow Cl + ClOO \qquad (7.42b)$$
$$ClOO + M \rightarrow Cl + O_2 + M \qquad (7.42c)$$
$$2Cl + 2O_3 \rightarrow 2ClO + 2O_2 \qquad (7.42d)$$

$$\text{Net:} \qquad 2O_3 + h\nu \rightarrow 3O_2 \qquad (7.43)$$

The following points should be noted about this reaction cycle:

1. Reactions (7.42) form a catalytic cycle in which ClO is the catalyst, because two ClO molecules are regenerated for every two ClO molecules that are consumed.
2. The cycle does not depend on atomic oxygen (which is in short supply).
3. The chlorine atom in the ClO on the left side of Reaction (7.42a) derives from Cl released from Freons via Reactions (7.34) and (7.35). However, as we have seen, the chlorine atom is then quickly tied up as HCl and ClONO$_2$ by Reactions (7.38) and (7.39). But, in the presence of PSCs, chlorine gas (Cl$_2$) is released by Reaction (7.40) and, as soon as the solar radiation reaches sufficient intensity in early spring, Reaction (7.41) releases atomic chlorine. Reaction (7.36a) converts this into ClO, which is then available for the first step in the reaction cycle (7.42) that leads to the rapid depletion of ozone in the Antarctic stratosphere.
4. The dimer (ClO)$_2$ is formed by Reaction (7.42a) only at low temperatures. Low-enough temperatures are present in the Antarctic stratosphere, where there are also large concentrations of ClO. Therefore, on both counts, the Antarctic stratosphere in spring is a region in which the reaction cycle (7.42) can destroy large quantities of ozone.

At this point, the reader might well ask if an ozone hole develops in the Arctic stratosphere in winter and, if not, why not? In fact, dramatic depletions of ozone have *not* been measured over the Arctic, although there is evidence for anomalous chlorine chemistry similar to that in the Antarctic. In a field study carried out in the Arctic in 1988–89, sharp increases were measured in the concentrations of ClO in the stratosphere, and these appeared to be associated with PSCs. Also, increases in OClO were measured, which provides support for the reaction cycle (7.42).[7] On the other hand, although some denitrification was measured at altitudes around 20 km, it was not as great as in the Antarctic stratosphere, perhaps because the PSCs evaporated in the lower stratosphere. Also, dehydration was much less in the Arctic. In any case, the decreases in total ozone column in the Arctic in 1988–89 were only a few percent, much less than observed in the Antarctic. It is not known whether this was due to insufficient anomalous chemistry or to less than optimal meteorological conditions for ozone depletion. For example, stratospheric temperatures remained very low until the middle of February in 1989, when there was a sudden warming and the PSCs disappeared. Thus, air that was sufficiently cold for Reaction (7.42a) to proceed rapidly may not have received sufficient solar radiation for Reaction (7.42b) to proceed effectively. In the Antarctic, stratospheric ozone is depleted primarily in September (which corresponds to March in the Arctic) when temperatures are still very low, but solar radiation is increasing rapidly. It would appear that while concentrations of Freons remain high in the atmosphere, the Arctic stratosphere has the potential to cause the same dramatic losses in ozone as the Antarctic stratosphere, but that the combination of chemical and meteorological conditions that lead to such reductions may not be as common in the Arctic as in the Antarctic.

On a global scale, ground-based and satellite observations show significant decreases of total column ozone at middle latitudes in the northern hemisphere of 2.7% per decade in winter, 1.3% decade in summer, and 1.2% per decade in the fall. Similar decreases are apparent at middle latitudes in the southern hemisphere; and at high latitudes, beneath the region of the Antarctic ozone hole, the decreases are 14% per year. The decreases have occurred primarily in the lower stratosphere. No trends in ozone concentrations have been observed in the tropics.

Concerns about the health and environmental hazards of increased

UV radiation at the Earth's surface, which accompany depletion in the total column ozone, have led to international agreements to reduce the manufacture of chlorofluorocarbons, and to eliminate them completely by the year 2000. Even so, due to the long lifetimes of Freons, the concentrations of chlorine in the stratosphere are expected to rise substantially. Even if the release into the atmosphere of all Freons were to stop right now, it would be well into the twenty-second century before atmospheric concentrations returned to the pre-1930s values. Thus, further decreases in total column ozone, and over increasingly larger regions of the globe, are to be expected.

Exercises

7.6.   Answer, interpret, or explain the following in light of the principles presented in this chapter.

(a) Photochemical reactions in the atmosphere generally involve UV and visible radiation.

(b) Photochemical processes are sometimes extremely efficient in converting light into chemical energy. Give an example.

(c) Ionization potentials are generally greater than corresponding binding energies.

(d) Astronauts see a faint envelope of light in the upper atmosphere on the night side of the Earth.

(e) If the primary photochemical quantum yields for a reaction are all small ($\ll 1$), the photophysical processes must be important.

(f) Assuming that the photostationary state given by Eq. (7.23) is strictly applicable in the atmosphere and that k_3 and $[O_3(g)]$ are approximately constant, how would you expect the ratio $[NO_2(g)]/[NO(g)]$ to vary diurnally in the atmosphere?

(g) In the troposphere air temperatures generally decrease with height, but in the stratosphere they start to increase with height near the level where ozone concentrations are a maximum.

(h) The use of nitrate fertilizers, which release nitrous oxide (N_2O) into the troposphere, could affect stratosphere ozone concentrations. (*Hint:* N_2O photodissociates for $\lambda < 0.25$ μm yielding NO.)

(i) Following the large volcanic eruption of Mt. Pinatubo in the Philippines in 1991, which injected large quantities of sulfur dioxide into the stratosphere, temporary depletions in stratospheric ozone of 5–8% were observed above the tropics.

(j) For catalytic cycles such as Reactions (7.28) to be efficient, each reaction comprising the cycle must be exothermic.

7.7. Compare the magnitude of the energy of an einstein of UV radiation (take $\lambda = 0.1$ μm) with the energies of a weak chemical bond (say 1 eV), a strong chemical bond (say 10 eV), and the kinetic energy of impact of two molecules (say 0.05 eV).

7.8. The shortest wavelength of solar radiation that reaches the lower atmosphere is about 0.3 μm. What is the energy associated with 1 einstein of this radiation?

7.9. The ionization energies for nitrogen (N_2) and oxygen (O_2) are 1,495 and 1,205 kJ mole^{-1}, respectively. What are the maximum wavelengths of solar radiation that can ionize nitrogen and oxygen? Where do these wavelengths lie in the em spectrum?

7.10. How many photons, with the required minimum energy, are required to produce 1.0 kg of OH in the following photolytic reaction?

$$H_2O(g) + h\nu \rightarrow OH(g) + H(g)$$

Assume that each photon is absorbed by a water molecule which then photolyses. The bonding energy of the O—H bond in H_2O is 5.2 eV.

7.11. If eight photons of red light of wavelength 0.685 μm are needed to produce one molecule of oxygen, and if the average energy stored in the photosynthetic process is 470 kJ per mole of oxygen, what is the energy conversion efficiency of photosynthesis with respect to oxygen?

7.12. The concentration of ozone just above the Earth's surface is 0.040 ppmv, and the rate coefficients for the reactions

$$NO_2(g) + h\nu \rightarrow NO(g) + O(g)$$

and,

$$O_3(g) + NO(g) \rightarrow NO_2(g) + O_2(g)$$

are $j = 4.0 \times 10^{-3}$ s^{-1} and $k_2 = 1.0 \times 10^{-20}$ m^3 molecule^{-1} s^{-1}, respectively. Use Eq. (7.23) to calculate the ratio of [NO_2(g)] to [NO(g)] at 20°C and 1 atm.

7.13. If n_1, n_2, n_3, and n_M are the concentrations of O, O_2, O_3, and the inert molecule M in reactions (7.24) to (7.27), write the expressions for dn_1/dt, dn_2/dt, and dn_3/dt in terms of n_1, n_2, n_3, and n_M and the four rate coefficients defined by Reactions (7.24) to (7.27).

7.14. Using the expressions for dn_1/dt, dn_2/dt, and dn_3/dt derived in Exercise 7.13, show that $n_1 + 2n_2 + 3n_3 =$ constant. Could you have predicted this result?

7.15. Write the catalytic cycles and the net reactions corresponding to Reaction (7.28) when X = H and when X = OH.

7.16. If the rate coefficient for Reaction (7.27) is fitted to the Arrhenius relation (3.9), the values of A and E_a are 8×10^{-12} cm^3 molecule^{-1} s^{-1} and 17.1×10^3 J mol^{-1}, respectively. The corresponding values of A and E_a for reaction (7.36b) are 4.7×10^{-11} cm^3 molecule^{-1} s^{-1} and 0.4×10^3 J mol^{-1}, respectively. Can the rate coefficients for the Reactions (7.27) and (7.36b) ever be the same? Which is the larger?

Notes

1 An electron-volt is the energy gained by one electron when it is accelerated through an electrical potential difference of 1 volt.

2 The term *radical* (also called *free radical*) refers to an atom or molecule containing an unpaired electron; as might be expected, radicals are very reactive.

3 In 1992 nitric oxide was voted "molecule of the year" by the journal *Science*. Not only is it a destroyer of ozone [see Reaction (7.22)], a suspected carcinogen, and a precursor of acid rain, it is essential to activities in the brain and in the body's immune system.

4 The percentage increase in DNA damage is about twice the percentage decrease in the total ozone column. "Total ozone column" is the integrated amount of ozone in a vertical column of unit horizontal cross-sectional area, extending from the Earth's surface to the top of the atmosphere. If all of the ozone in such a column were to be brought to STP, it would have a depth of only a few millimeters, and most of this would derive from the stratosphere.

5 In the remainder of this chapter, all of the chemical species are gases unless stated otherwise. Therefore, for conciseness, we will drop the parenthetical "g," indicating gas following the chemical symbol.

6 This account is based in part on a very readable discussion of this topic given in *Chemistry of Atmospheres* by Richard P. Wayne, Oxford University Press (1991), to which the reader is referred for more detailed information and original references.

7 The symmetric form of chlorine dioxide (OClO) is different from the unstable species ClOO in Reaction (7.42b). However, the presence of OClO provides an important indication of the amount of ClO and therefore the destruction of odd oxygen in the ozone hole.

Appendix I. *International system of units (SI)[a]*

Quantity	Name of unit	Symbol	Definition
	Basic units		
Length	meter	m	
Mass	kilogram	kg	
Time	second	s	
Electrical current	ampere	A	
Temperature	degree Kelvin	K	
	Derived units		
Force	newton	N	$kg\ m\ s^{-2}$
Pressure	pascal	Pa	$N\ m^{-2} = kg\ m^{-1}\ s^{-2}$
Energy	joule	J	$kg\ m^2\ s^{-2}$
Power	watt	W	$J\ s^{-1} = kg\ m^2\ s^{-3}$
Electric potential difference	volt	V	$W\ A^{-1} = kg\ m^2\ s^{-3}\ A^{-1}$
Electrical charge	coulomb	C	$A\ s$
Electrical resistance	ohm	Ω	$V\ A^{-1} = kg\ m^2\ s^{-3}\ A^{-2}$
Electrical capacitance	farad	F	$A\ S\ V^{-1} = kg^{-1}\ m^2\ s^4\ A^2$
Frequency	hertz	Hz	s^{-1}
Celsius temperature	degree Celsius	°C	$K - 273.15$
Temperature interval	degree	degree or °	K or °C need not be specified

[a] The SI system of units is the internationally accepted form of the metric system. The SI system is being used increasingly in chemistry, as in other sciences, but some older units persist. This book reflects this dichotomy, although the SI system has been used as much as possible. Some useful conversion factors between various units are given in Appendix II.

Prefixes used to construct decimal multiples of units

Multiple	Prefix	Symbol	Multiple	Prefix	Symbol
10^{-1}	deci	d	10	deca	da
10^{-2}	centi	c	10^2	hecto	h
10^{-3}	milli	m	10^3	kilo	k
10^{-6}	micro	μ	10^6	mega	M
10^{-9}	nano	n	10^9	giga	G
10^{-12}	pico	p	10^{12}	tera	T
10^{-15}	femto	f	10^{15}	peta	P
10^{-18}	atto	a	10^{18}	exa	E

Universal constants

Universal gas constant – in SI units (R^*)	8.3143 J deg^{-1} mol^{-1}
Universal gas constant – in "chemical units" (R_c^*)	0.0821 L atm deg^{-1} mol^{-1}
Boltzmann constant (k)	1.381×10^{-23} J deg^{-1} molecule^{-1}
Avogadro's number (N_A)	6.022×10^{23} molecules mol^{-1}
Faraday constant (F)	$96{,}489$ C equiv^{-1}
Planck constant (h)	6.6262×10^{-34} J s
Velocity of light in a vacuum (c)	2.998×10^8 m s^{-1}

Other values

Number of molecules in 1 m^3 of a gas at 1 atm and 0°C (Loschmidt number)	2.69×10^{25} molecules m^{-3}
Volume of 1 mole of an ideal gas at 0°C and 1 atm	22.415 L
Ion-product constant for water at 25°C and 1 atm (K_w)	1.00×10^{-14} mol^2 L^{-2}

Conversion factors

1 bar $= 10^5$ Pa

1 atm $= 1.013$ bar $= 760.0$ Torr

1 ppm[a] $= 2.46 \times 10^{19}$ molecules m$^{-3} = 40.9$ (molecular weight of species) μg m^{-3}

For a second-order rate coefficient: 1 cm^3 s^{-1} molecule$^{-1} = 6.02 \times 10^{20}$ liter s^{-1} mol^{-1}

For a third-order rate coefficient: 1 cm^6 s^{-1} molecule$^{-2} = 3.63 \times 10^{41}$ liter^2 s^{-1} mol^{-2}

1 kcal mol$^{-1} = 4.18$ kJ mol^{-1}

1 eV $= 96.489$ kJ mol^{-1}

$\ln x = 2.3026 \log x$

---

[a] For a gas *by mass* relative to air at 1 atm and 25°C. *Note:* 1 ppb $= 10^{-3}$ ppm and 1 ppt $= 10^{-6}$ ppm.

Appendix III. *Atomic weights*[a]

Element	Symbol	Atomic weight
Actinium	Ac	(227)
Aluminum	Al	26.9815
Americium	Am	(243)
Antimony	Sb	121.75
Argon	Ar	39.948
Arsenic	As	74.9216
Astatine	At	(210)
Barium	Ba	137.34
Berkelium	Bk	(247)
Beryllium	Be	9.0122
Bismuth	Bi	208.980
Boron	B	10.811
Bromine	Br	79.904
Cadmium	Cd	112.40
Calcium	Ca	40.08
Californium	Cf	(251)
Carbon	C	12.01115
Cerium	Ce	140.12
Cesium	Cs	132.905
Chlorine	Cl	35.453
Chromium	Cr	51.996
Cobalt	Co	58.9332
Copper	Cu	63.546
Curium	Cm	(247)
Dysprosium	Dy	162.50
Einsteinium	Es	(252)
Erbium	Er	167.26
Europium	Eu	151.96
Fermium	Fm	(257)
Fluorine	F	18.9984
Francium	Fr	(223)
Gadolinium	Gd	157.25
Gallium	Ga	69.72
Germanium	Ge	72.59
Gold	Au	196.967
Hafnium	Hf	178.49
Helium	He	4.0026
Holmium	Ho	164.930
Hydrogen	H	1.00794
Indium	In	114.82
Iodine	I	126.904
Iridium	Ir	192.2
Iron	Fe	55.847
Krypton	Kr	83.80
Lanthanum	La	138.91
Lawrencium	Lw	(260)
Lead	Pb	207.19

Appendix III *(Cont.)*

Element	Symbol	Atomic weight
Lithium	Li	6.939
Lutetium	Lu	174.97
Magnesium	Mg	24.305
Manganese	Mn	54.938
Mendelevium	Md	(258)
Mercury	Hg	200.59
Molybdenum	Mo	95.94
Neodymium	Nd	144.24
Neon	Ne	20.179
Neptunium	Np	(237)
Nickel	Ni	58.69
Niobium	Nb	92.906
Nitrogen	N	14.0067
Nobelium	No	(259)
Osmium	Os	190.2
Oxygen	O	15.9994
Palladium	Pd	106.4
Phosphorus	P	30.9738
Platinum	Pt	195.09
Plutonium	Pu	(244)
Polonium	Po	(209)
Potassium	K	39.1
Praseodymium	Pr	140.907
Promethium	Pm	(145)
Protactinium	Pa	(231)
Radium	Ra	(226)
Radon	Rn	(222)
Rhenium	Re	186.2
Rhodium	Rh	102.905
Rubidium	Rb	85.47
Rutherfordium	Rf	(261)
Ruthenium	Ru	101.07
Samarium	Sm	150.35
Scandium	Sc	44.956
Selenium	Se	78.96
Silicon	Si	28.086
Silver	Ag	107.868
Sodium	Na	22.9898
Strontium	Sr	87.62
Sulfur	S	32.066
Tantalum	Ta	180.948
Technetium	Tc	(98)
Tellurium	Te	127.60
Terbium	Tb	158.924
Thallium	Tl	204.38
Thorium	Th	232.038
Thulium	Tm	168.934

Appendix III *(Cont.)*

Element	Symbol	Atomic weight
Tin	Sn	118.71
Titanium	Ti	47.88
Tungsten	W	183.85
Uranium	U	238.03
Vanadium	V	50.942
Xenon	Xe	131.29
Ytterbium	Yb	173.04
Yttrium	Y	88.906
Zinc	Zn	65.39
Zirconium	Zr	91.22

[a] Based on an atomic weight for C^{12} of 12.000 .... Values in parentheses are for the most stable known isotopes.

Appendix IV. Equilibrium (or dissociation) constants for some chemical reactions
(a) Acids in pure water at 25°C and 1 atm[a]

Acid	Strength	Reaction	K_a
Hydrochloric acid	Very strong	$HCl + H_2O \rightleftarrows H_3O^+ + Cl^-$	10^7
Sulfuric acid		$H_2SO_4 + H_2O \rightleftarrows H_3O^+ + HSO_4^-$	10^3
Nitric acid		$HNO_3 + H_2O \rightleftarrows H_3O^+ + NO_3^-$	10
Sulfurous acid ($SO_2 + H_2O$)	Strong	$H_2SO_3 + H_2O \rightleftarrows H_3O^+ + HSO_3^-$	1.7×10^{-2}
Hydrogen sulfate ion		$HSO_4^- + H_2O \rightleftarrows H_3O^+ + SO_4^{2-}$	1.3×10^{-2}
Phosphoric acid		$H_3PO_4 + H_2O \rightleftarrows H_3O^+ + H_2PO_4^-$	7.1×10^{-3}
Ferric ion		$Fe(H_2O)_6^{3+} + H_2O \rightleftarrows H_3O^+ + Fe(H_2O)_5OH^{2+}$	6.0×10^{-3}
Hydrofluoric acid	Weak	$HF + H_2O \rightleftarrows H_3O^+ + F^-$	6.7×10^{-4}
Nitrous acid		$HNO_2 + H_2O \rightleftarrows H_3O^+ + NO_2^-$	5.1×10^{-4}
Formic acid		$HCOOH + H_2O \rightleftarrows H_3O^+ + HCOO^-$	1.8×10^{-4}
Chromic ion		$Cr(H_2O)_6^{3+} + H_2O \rightleftarrows H_3O^+ + Cr(H_2O)_5OH^{2+}$	1.5×10^{-4}
Acetic acid	Weak	$CH_3COOH + H_2O \rightleftarrows H_3O^+ + CH_3COO^-$	1.8×10^{-5}
Aluminum ion		$Al(H_2O)_6^{3+} + H_2O \rightleftarrows H_3O^+ + Al(H_2O)_5OH^{2+}$	1.4×10^{-5}

			K_a
Carbonic acid ($CO_2 + H_2O$)		$H_2CO_3 + H_2O \rightleftarrows H_3O^+ + HCO_3^-$	4.4×10^{-7}
Hydrogen sulfide		$H_2S + H_2O \rightleftarrows H_3O^+ + HS^-$	1.0×10^{-7}
Dihydrogen phosphate ion		$H_2PO_4^- + H_2O \rightleftarrows H_3O^+ + HPO_4^{2-}$	6.3×10^{-8}
Hydrogen sulfite ion		$HSO_3^- + H_2O \rightleftarrows H_3O^+ + SO_3^{2-}$	6.2×10^{-8}
Ammonium ion	Weak	$NH_4^+ + H_2O \rightleftarrows H_3O^+ + NH_3$	5.7×10^{-10}
Hydrogen carbonate ion		$HCO_3^- + H_2O \rightleftarrows H_3O^+ + CO_3^{2-}$	4.7×10^{-11}
Hydrogen peroxide	Very weak	$H_2O_2 + H_2O \rightleftarrows H_3O^+ + HO_2^-$	2.4×10^{-12}
Monohydrogen phosphate ion		$HPO_4^{2-} + H_2O \rightleftarrows H_3O^+ + PO_4^{3-}$	4.4×10^{-13}
Hydrogen sulfide ion		$HS^- + H_2O \rightleftarrows H_3O^+ + S^{2-}$	1.3×10^{-13}
Water		$H_2O + H_2O \rightleftarrows H_3O^+ + OH^-$	1.0×10^{-14}
Hydroxide ion		$OH^- + H_2O \rightleftarrows H_3O^+ + O^{2-}$	$<10^{-36}$

[a] The reactions are of the form $HA + H_2O(l) \rightleftarrows H_3O^+(aq) + A^-(aq)$. The *equilibrium* (or *acid-dissociation* or *ioniza-tion*) constant is defined as

$$K_c = K_a = \frac{[H_3O^+(aq)][A^-(aq)]}{[HA][H_2O(l)]}$$

where $H_2O(l) = 1$. The equilibrium constant is often expressed in an analogous way to that used for pH, namely, $pK_a = -\log K_a$. The base-dissociation constant K_b for the conjugate base, $A^-(aq)$, of HA is given by $K_b = 1.00 \times 10^{-14}/K_a$ at 25°C; see Eq. (5.13).

(b) Successive acid-dissociation constants of some polyprotic acids at 25°C and 1 atm
(see Section 5.7)

Acid	Reaction	Acid-dissociation constant
Carbonic[a]	$H_2CO_3 + H_2O \rightleftarrows H_3O^+ + HCO_3^-$ $HCO_3^- + H_2O \rightleftarrows H_3O^+ + CO_3^{2-}$	$K_{a_1} = 4.4 \times 10^{-7}$ $K_{a_2} = 4.7 \times 10^{-11}$
Hydrosulfuric[b]	$H_2S + H_2O \rightleftarrows H_3O^+ + HS^-$ $HS^- + H_2O \rightleftarrows H_3O^+ + S^{2-}$	$K_{a_1} = 1.0 \times 10^{-7}$ $K_{a_2} = 1 \times 10^{-19}$
Oxalic	$H_2C_2O_4 + H_2O \rightleftarrows H_3O^+ + HC_2O_4^-$ $HC_2O_4^- + H_2O \rightleftarrows H_3O^+ + C_2O_4^-$	$K_{a_1} = 5.4 \times 10^{-2}$ $K_{a_2} = 5.4 \times 10^{-5}$
Phosphoric	$H_3PO_4 + H_2O \rightleftarrows H_3O^+ + H_2PO_4^-$ $H_2PO_4^- + H_2O \rightleftarrows H_3O^+ + HPO_4^{2-}$ $HPO_4^{2-} + H_2O \rightleftarrows H_3O^+ + PO_4^{3-}$	$K_{a_1} = 7.1 \times 10^{-3}$ $K_{a_2} = 6.3 \times 10^{-8}$ $K_{a_3} = 4.2 \times 10^{-13}$

Phosphorous	$H_3PO_3 + H_2O \rightleftharpoons H_3O^+ + H_2PO_3^-$	$K_{a1} = 3.7 \times 10^{-2}$
	$H_2PO_3^- + H_2O \rightleftharpoons H_3O^+ + HPO_3^{2-}$	$K_{a2} = 2.1 \times 10^{-7}$
Sulfurous[c]	$H_2SO_3 + H_2O \rightleftharpoons H_3O^+ + HSO_3^-$	$K_{a1} = 1.3 \times 10^{-2}$
	$HSO_3^- + H_2O \rightleftharpoons H_3O^+ + SO_3^{2-}$	$K_{a2} = 6.2 \times 10^{-8}$
Sulfuric[d]	$H_2SO_4 + H_2O \rightleftharpoons H_3O^+ + HSO_4^-$	$K_{a1} = $ very large
	$HSO_4^- + H_2O \rightleftharpoons H_3O^+ + SO_4^{2-}$	$K_{a2} = 1.1 \times 10^{-2}$

[a] H_2CO_3 cannot be isolated. It is in equilibrium with H_2O and dissolved CO_2. The value given for K_{a1} is actually for the reaction

$$CO_2(aq) + 2\ H_2O \rightleftharpoons H_3O^+ + HCO_3^-$$

Generally, aqueous solutions of CO_2 are treated *as if* the $CO_2(aq)$ were first converted to H_2CO_3, followed by ionization of the H_2CO_3.

[b] The value for K_{a2} of H_2S most commonly found in the older literature is about 1×10^{-14}, but current evidence suggests that the best value is considerably smaller.

[c] H_2SO_3 is a hypothetical, nonisolatable species. The value listed for K_{a1} is actually for the reaction

$$SO_2(aq) + 2\ H_2O \rightleftharpoons H_3O^+ + HSO_3^-$$

[d] H_2SO_4 is completely ionized in the first step.

(c) Solubility (or ion) product constants for some salts at 25°C and 1 atm[a]

Salt	Solubility equilibrium	K_{sp}
Aluminum hydroxide	$Al(OH)_3(s) \rightleftarrows Al^{3+}(aq) + 3OH^-(aq)$	1.3×10^{-33}
Barium carbonate	$BaCO_3(s) \rightleftarrows Ba^{3+}(aq) + CO_3^{2-}(aq)$	5.1×10^{-9}
Barium hydroxide	$Ba(OH)_2(s) \rightleftarrows Ba^{2+}(aq) + 2OH^-(aq)$	5×10^{-3}
Barium sulfate	$BaSO_4(s) \rightleftarrows Ba^{2+}(aq) + SO_4^{2-}(aq)$	1.1×10^{-10}
Bismuth(III) sulfide	$Bi_2S_3(s) \rightleftarrows 2Bi^{3+}(aq) + 3S^{2-}(aq)$	1×10^{-97}
Cadmium sulfide	$CdS(s) \rightleftarrows Cd^{2+}(aq) + 3S^{2-}(aq)$	8.0×10^{-28}
Calcium carbonate	$CaCO_3(s) \rightleftarrows Ca^{2+}(aq) + CO_3^{2-}(aq)$	2.8×10^{-9}
Calcium fluoride	$CaF_2(s) \rightleftarrows Ca^{2+}(aq) + 2F^-(aq)$	5.3×10^{-9}
Calcium hydroxide	$Ca(OH)_2(s) \rightleftarrows Ca^{2+}(aq) + 2OH^-(aq)$	5.5×10^{-6}
Calcium sulfate	$CaSO_4(s) \rightleftarrows Ca^{2+}(aq) + SO_4^{2-}(aq)$	9.1×10^{-6}
Chromium(III) hydroxide	$Cr(OH)_3(s) \rightleftarrows Cr^{3+}(aq) + 3OH^-(aq)$	6.3×10^{-31}
Cobalt(II) sulfide	$CoS(s) \rightleftarrows Co^{2+}(aq) + S^{2-}(aq)$	4.0×10^{-21}
Copper(II) sulfide	$CuS(s) \rightleftarrows Cu^{2+}(aq) + S^{2-}(aq)$	6×10^{-37}
Copper carbonate	$CuCO_3(s) \rightleftarrows Cu^{2+}(aq) + CO_3^{2-}(aq)$	1.4×10^{-10}
Iron(II) sulfide	$FeS(s) \rightleftarrows Fe^{2+}(aq) + S^{2-}(aq)$	6×10^{-19}
Iron(III) hydroxide	$Fe(OH)_3(s) \rightleftarrows Fe^{3+}(aq) + 3OH^-(aq)$	4×10^{-38}
Lead(II) chloride	$PbCl_2(s) \rightleftarrows Pb^{2+}(aq) + 2Cl^-(aq)$	1.6×10^{-5}
Lead(II) chromate	$PbCrO_4(s) \rightleftarrows Pb^{2+}(aq) + CrO_4^{2-}(aq)$	2.8×10^{-13}
Lead(II) iodide	$PbI_2(s) \rightleftarrows Pb^{2+}(aq) + 2I^-(aq)$	7.1×10^{-9}
Lead(II) sulfate	$PbSO_4(s) \rightleftarrows Pb^{2+}(aq) + SO_4^{2-}(aq)$	1.6×10^{-8}
Lead(II) sulfide	$PbS(s) \rightleftarrows Pb^{2+}(aq) + S^{2-}(aq)$	3.0×10^{-28}
Lithium phosphate	$Li_3PO_4(s) \rightleftarrows 3Li^+(aq) + PO_4^{3-}(aq)$	3.2×10^{-9}

Magnesium carbonate	$MgCO_3(s) \rightleftharpoons Mg^{2+}(aq) - CO_3^{2-}(aq)$	3.5×10^{-8}
Magnesium fluoride	$MgF_2(s) \rightleftharpoons Mg^{2+}(aq) - 2F^-(aq)$	3.7×10^{-8}
Magnesium hydroxide	$Mg(OH)_2(s) \rightleftharpoons Mg^{2+}(aq) - 2OH^-(aq)$	1.9×10^{-13}
Magnesium phosphate	$Mg_3(PO_4)_2(s) \rightleftharpoons 3Mg^{2+}(aq) + 2PO_4^{3-}(aq)$	1×10^{-25}
Manganese(II) sulfide	$MnS(s) \rightleftharpoons Mn^{2+}(aq) + S^{2-}(aq)$	3×10^{-14}
Mercury(I) chloride	$Hg_2Cl_2(s) \rightleftharpoons Hg_2^{2+}(aq) + 2Cl^-(aq)$	1.3×10^{-18}
Mercury(II) sulfide	$HgS(s) \rightleftharpoons Hg^{2+}(aq) + S^{2-}(aq)$	2×10^{-53}
Nickel(II) sulfide	$NiS(s) \rightleftharpoons Ni^{2+}(aq) + S^{2-}(aq)$	3.2×10^{-19}
Silver bromide	$AgBr(s) \rightleftharpoons Ag^+(aq) + Br^-(aq)$	5.0×10^{-13}
Silver carbonate	$Ag_2CO_3(s) \rightleftharpoons 2Ag^+(aq) + CO_3^{2-}(aq)$	8.1×10^{-12}
Silver chloride	$AgCl(s) \rightleftharpoons Ag^+(aq) + Cl^-(aq)$	1.8×10^{-10}
Silver chromate	$Ag_2CrO_4(s) \rightleftharpoons 2Ag^+(aq) + CrO_4^{2-}(aq)$	2.4×10^{-12}
Silver iodide	$AgI(s) \rightleftharpoons Ag^+(aq) + I^-(aq)$	8.5×10^{-17}
Silver sulfate	$Ag_2SO_4(s) \rightleftharpoons 2Ag^+(aq) + SO_4^{2-}(aq)$	1.4×10^{-5}
Silver sulfide	$Ag_2S(s) \rightleftharpoons 2Ag^+(aq) + S^{2-}(aq)$	6×10^{-51}
Strontium carbonate	$SrCO_3(s) \rightleftharpoons Sr^{2-}(aq) + CO_3^{2-}(aq)$	1.1×10^{-10}
Strontium sulfate	$SrSO_4(s) \rightleftharpoons Sr^{2-}(aq) + SO_4^{2-}(aq)$	3.2×10^{-7}
Tin(II) sulfide	$SnS(s) \rightleftharpoons Sn^{2-}(aq) + S^{2-}(aq)$	1.0×10^{-26}
Zinc sulfide	$ZnS(s) \rightleftharpoons Zn^{2+}(aq) + S^{2-}(aq)$	2×10^{-25}

[a]The solubility (or ion) product constant (K_{sp}) is defined in the same way as the equilibrium (or dissociation) constant – see Section 4.6.

(d) *Formation constants (K_f) for some complex ions. All of the ions are in aqueous solutions at 25°C and 1 atm. Some of the data are quite uncertain[a]*

Complex ion	Equilibrium reaction	K_f
$AgCl_2^-$	$AgCl(s) + Cl^- \rightleftarrows AgCl_2^-$	2.0×10^{-5}
$CuCl_2^-$	$CuCl(s) + Cl^- \rightleftarrows CuCl_2^-$	1.0×10^{-1}
$CuCl_3^{2-}$	$CuCl_2^- + Cl^- \rightleftarrows CuCl_3^{2-}$	5×10^5
HgS_2^{2-}	$HgS(s) + S^{2-} \rightleftarrows HgS_2^{2-}$	4.0×10^0
$ZnCl^+$	$Zn^{2+} + Cl^- \rightleftarrows ZnCl^+$	1.0×10^1
$CuCl^+$	$Cu^{2+} + Cl^- \rightleftarrows CuCl^+$	1.0×10^1
ZnF^+	$Zn^{2+} + F^- \rightleftarrows ZnF^+$	2.0×10^1
$PbCl^+$	$Pb^{2+} + Cl^- \rightleftarrows PbCl^+$	3.0×10^1
$FeCl^{2+}$	$Fe^{3+} + Cl^- \rightleftarrows FeCl^{2+}$	3.0×10^1
$SnCl^+$	$Sn^{2+} + Cl^- \rightleftarrows SnCl^+$	6.3×10^1
$\{Cu(CO_3)_2\}^{2-}$	$CuCO_3(aq) + CO_3^{2-} \rightleftarrows \{Cu(CO_3)_2\}^{2-}$	1.6×10^3
SnF^+	$Sn^{2+} + Cl^- \rightleftarrows SnF^+$	7.7×10^4
SnS_3^{2-}	$SnS_2(s) + S^{2-} \rightleftarrows SnS_3^{2-}$	1.0×10^5
FeF^{2+}	$Fe^{3+} + F^- \rightleftarrows FeF^{2+}$	1.6×10^5
AlF^{2+}	$Al^{3+} + F^- \rightleftarrows AlF^{2+}$	1.6×10^6
$\{Ag(NH_3)_2\}^+$	$Ag^+ + 2NH_3(aq) \rightleftarrows \{Ag(NH_3)_2\}^+$	1.6×10^7
$\{Ni(NH_3)_6\}^{2+}$	$Ni^{2+} + 6NH_3 \rightleftarrows \{Ni(NH_3)_6\}^{2+}$	5.5×10^8
$\{Cu(NH_3)_4\}^{2+}$	$Cu^{2+} + 4NH_3 \rightleftarrows \{Cu(NH_3)_4\}^{2+}$	1.1×10^{13}
$\{Ag(S_2O_3)_2\}^{3-}$	$Ag^+ + 2S_2O_3^{2-} \rightleftarrows \{Ag(S_2O_3)_2\}^{3-}$	1.7×10^{13}
$\{Ag(CN)_2\}^-$	$Ag^+ + 2CN^- \rightleftarrows \{Ag(CN)_2\}^-$	5.6×10^{18}
$\{Cu(CN)_4\}^{2-}$	$Cu^{2+} + 4CN^- \rightleftarrows \{Cu(CN)_4\}^{2-}$	1.0×10^{25}
$\{Fe(CN)_6\}^{4-}$	$Fe^{2+} + 6CN^- \rightleftarrows \{Fe(CN)_6\}^{4-}$	10^{37}
$\{Fe(CN)_6\}^{3-}$	$Fe^{3+} + 6CN^- \rightleftarrows \{Fe(CN)_6\}^{3-}$	1.0×10^{42}

[a]The equilibrium constant for the formation of a complex ion from a metal ion in aqueous solution is called the formation constant (K_f) of the complex ion. The higher the value of K_f the more stable is the complex

Appendix V. *Some molar standard Gibbs free energies of formation* $(\Delta \overline{G}_f^0)$, *molar standard enthalpies (or heats) of formation* $(\Delta \overline{H}_f^0)$ *and molar absolute entropies* $(\overline{S}^0)$ *at 25°C and 1 atmosphere*[a]

Chemical formula[b]	Name	$\Delta \overline{G}_f^0$ (kJ mol^{-1})	$\Delta \overline{H}_f^0$ (kJ mol^{-1})	$\overline{S}^0$ (J deg^{-1} mol^{-1})
Ca(s)	Calcium	0	0	41.6
CaCO$_3$(s)	Calcite	−1129	−1207	88.7
CaSO$_4$(s)	Calcium sulphate	−1322	−1434	106.7
C(g)	Carbon	671.3	716.7	158.0
C(s)	Graphite	0	0	5.7
C(s)	Diamond	2.9	1.9	2.4
CH$_4$(g)	Methane	−50.8	−74.8	186.2
C$_2$H$_6$(g)	Ethane	−32.9	−84.7	229.5
CO(g)	Carbon monoxide	−137.2	−110.5	197.6
CO$_2$(g)	Carbon dioxide	−394.4	−393.5	213.7
CO$_2$(aq)	Carbon dioxide	−386.2	−412.9	121.3
Cl$_2$(g)	Chlorine	0	0	223.0
HCl(g)	Hydrochloric acid	−95.3	−92.3	186.8
Cl$^-$(g)	Chloride ion	−131.2	−167.4	56.5
Cu(s)	Copper	0	0	33.15
Cu$_2$O(s)	Cuprite	−148.5	−170.7	93.7
CuS(s)	Covellite	−51.0	−50.6	66.5
Cu$^+$(aq)	Cuprous ion or copper (I)	51.0	71.5	36.8
Cu^{2+}(aq)	Cupric ion or copper (II)	65.52	64.77	−99.6
F$_2$(g)	Fluorine	0	0	202.9
HF(g)	Hydrofluoric acid	−273.7	−271.1	173.5
HF(aq)	Hydrofluoric acid	−294.6	−329.3	108.8
F$^-$(aq)	Fluoride ion	−276.5	−329.1	−9.6
Au(s)	Gold	0	0	47.3
AuCl$_3$	Gold chloride	−45.2	−115.1	148.1
H$_2$(g)	Hydrogen	0	0	130.6
H$_2$O(g)	Water vapor	−228.6	−241.8	188.7
H$_2$O(l)	Water	−237.2	−285.6	69.9
H$_2$O$_2$(g)	Hydrogen peroxide	−105.5	−136.1	239.9
H$_2$O$_2$(l)	Hydrogen peroxide	−120.4	−187.6	110
H$_2$O$_2$(aq)	Hydrogen peroxide	......	−191.1	......
Fe(s)	Iron	0	0	27.2
Fe$_3$O$_4$(s)	Magnetite	−1015	−1118	146.4
FeS$_2$(s)	Pyrite	−160.2	−171.5	52.9
Fe^{2+}(aq)	Ferrous ion or iron (II)	−78.87	−89.1	−137.7
Fe^{3+}(aq)	Ferric ion or iron (III)	−4.6	−48.5	−316

Appendix V *(Cont.)*

Chemical formula[b]	Name	$\Delta \bar{G}_f^0$ (kJ mol^{-1})	$\Delta \bar{H}_f^0$ (kJ mol^{-1})	$\bar{S}^0$ (J deg^{-1} mol^{-1})
Pb(s)	Lead	0	0	64.8
Pb(g)	Lead	161.1	193.7	175.3
PbO(g)	Lead monoxide gas	20.9	42.3	239.7
Pb^{2+}(aq)	Lead ion	-24.4	-2	11
Mg(s)	Magnesium	0	0	32.6
MgO(s)	Periclase	-569.4	-601.7	26.9
Mg^{2+}(aq)	Magnesium ion	-455.2	-466.6	-138
Mn(s)	Manganese	0	0	32.0
MnO$_2$(s)	Pyrolusite	-465.1	-520.0	53.1
Mn^{2+}(aq)	Manganous ion	-227.6	-220.7	-73.6
Hg(l)	Mercury	0	0	76.0
Hg(g)	Mercury gas	31.85	61.5	174.9
HgCl(s)	Calomel	-104.6	-131.8	96.2
HgS(s)	Cinnabar	-50.2	-58.2	81.6
Hg$_2^{2+}$(aq)	Mercurous ion or mercury (I)	152.3	167.8	74.1
Hg^{2+}(aq)	Mercuric ion or mercury (II)	164.4	174.1	-22.6
N$_2$(g)	Nitrogen	0	0	191.5
NO(g)	Nitrogen monoxide	86.6	90.4	210.6
NO$_2$(g)	Nitrogen dioxide	51.8	33.9	240.5
N$_2$O(g)	Nitrous oxide	103.6	81.5	219.9
NH$_3$(g)	Ammonia	-16.5	-46.0	192.3
NH$_4$OH (aq)	Ammonium hydroxide	-263.6	-366.5	179.1
NO$_3^-$(aq)	Nitrate ion	-108.7	-205.0	146.4
NH$_4^+$(aq)	Ammonium	-79.5	-132.6	113.4
O$_2$(g)	Oxygen	0	0	205.0
O$_3$(g)	Ozone	163.2	142.7	239
OH$^-$(aq)	Hydroxide ion	-157.3	-230.0	-10.5
K(s)	Potassium	0	0	64.2
KCl(s)	Sylvite	-407.9	-436.0	82.4
K$^+$(aq)	Potassium ion	-281.6	-251.0	101.3
Si(s)	Silicon	0	0	18.9
SiO$_2$(s)	α-Quartz	-856.7	-910.9	41.8
Ag(s)	Silver	0	0	42.7
AgCl(s)	Silver chloride	-109.6	-126.8	96.2
Ag$^+$(aq)	Silver ion	77.0	-105.9	72.7
Na(s)	Sodium	0	0	51.2
NaCl(s)	Salt (or halite)	-384.1	-410.9	72.5
Na$^+$(aq)	Sodium ion	-261.5	-240.0	58.6

Appendix V *(Cont.)*

Chemical formula[b]	Name	$\Delta \overline{G}_f^0$ (kJ mol^{-1})	$\Delta \overline{H}_f^0$ (kJ mol^{-1})	$\overline{S}^0$ (J deg^{-1} mol^{-1})
S(s)	Sulfur (rhombic)	0	0	31.9
S$_2$(g)	Sulfur gas	80.8	130.0	228.0
H$_2$S(g)	Hydrogen sulfide	-33.6	-20.6	205.7
H$_2$S(aq)	Hydrogen sulfide	-27.4	-39.3	123.0
HS$^-$(aq)	Hydrogen sulfide ion (or bisulfide)	12.6	-17.2	62.8
S^{2-}(aq)	Sulfide	86.2	32.6	-16.7
SO$_2$(g)	Sulfur dioxide	-300.2	-296.9	248.1
SO$_3$(g)	Sulfur trioxide	-369.9	-395.0	256.1
SO$_4^{2-}$(aq)	Sulfate ion	-742.0	-907.5	20
HSO$_3^-$(aq)	Hydrogen sulfite ion (or bisulfite)	-527.73	-626.22	139.7
HSO$_4^-$(aq)	Hydrogen sulfate ion (or bisulfate)	-752.9	-885.8	126.8
U(s)	Uranium	0	0	50.3
UO$_2$(s)	Uraninite	-1031.8	-1084.5	77.9
Zn(s)	Zinc	0	0	41.6
Zn(g)	Zinc gas	95.0	130.5	14.2
ZnO(s)	Zinc oxide	-318.3	-348.3	43.64
Zn^{2+}(aq)	Zinc ion	-147.3	-152.3	-112

[a] The numbers in the table have different accuracies, and may change as new data becomes available.
    The molar standard free energy change for a chemical reaction ($\Delta \overline{G}^0$) is given by Eq. (2.34). The molar standard enthalpy change for a chemical reaction ($\Delta \overline{H}_{rx}^0$) is given by Eq. (2.14). The change in the molar absolute entropy for a chemical reaction ($\Delta \overline{S}^0$) is given by Eq. (2.24). To obtain the molar Gibbs free energy change for a chemical reaction at temperature T (i.e., $\Delta \overline{G}$) use the following approximation to the Gibbs–Helmholtz equation: $\Delta \overline{G} \approx \Delta \overline{H}_{rx}^0 - T\Delta \overline{S}^0$, which assumes that $\Delta \overline{H}_{rx}$ and $\Delta \overline{S}$ do not differ greatly from $\Delta \overline{H}_{rx}^0$ and $\Delta \overline{S}^0$, respectively. The equilibrium constant for a chemical reaction under standard conditions is related to $\Delta \overline{G}^0$ by Eq. (2.44), and at temperature T it is related to $\Delta \overline{G}_t^0$ by Eq. (2.42).
[b] s = solid; l = liquid; g = gas; aq = dissolved in water at a concentration of 1 M.

Appendix VI. *Names, formulas, and charges of some common ions*[a]

Positive ions (cations)		Negative ions (anions)	
Aluminum	Al^{3+}	Acetate	CH_3COO^-
Ammonium	NH_4^+	Bromide	Br^-
Barium	Ba^{2+}	Carbonate	CO_3^{2-}
Calcium	Ca^{2+}	Hydrogen carbonate	HCO_3^-
Chromium (II), chromous	Cr^{2+}	ion, bicarbonate	
Chromium (III), chromic	Cr^{3+}	Chlorate	ClO_3^-
Copper (I), cuprous	Cu^+	Chloride	Cl^-
Copper (II), cupric	Cu^{2+}	Chlorite	ClO_2^-
Hydrogen, hydronium	H^+, H_3O^+	Chromate	CrO_4^{2-}
Iron (II), ferrous	Fe^{2+}	Dichromate	$Cr_2O_7^{2-}$
Iron (III), ferric	Fe^{3+}	Fluoride	F^-
Lead	Pb^{2+}	Hydroxide	OH^-
Lithium	Li^+	Hypochlorite	ClO^-
Magnesium	Mg^{2+}	Iodide	I^-
Manganese (II),	Mn^{2+}	Nitrate	NO_3^-
manganous		Nitrite	NO_2^-
Mercury (I), mercurous	Hg_2^{2+}	Oxalate	$C_2O_4^{2-}$
Mercury (II), mercuric	Hg^{2+}	Hydrogen oxalate	$HC_2O_4^-$
Potassium	K^+	ion, binoxalate	
Silver	Ag^+	Perchlorate	ClO_4^-
Sodium	Na^+	Permanganate	MnO_4^-
Tin (II), stannous	Sn^{2+}	Phosphate	PO_4^{3-}
Tin (IV), stannic	Sn^{4+}	Monohydrogen	HPO_4^{2-}
Zinc	Zn^{2+}	phosphate	
		Dihydrogen	$H_2PO_4^-$
		phosphate	
		Sulfate	SO_4^{2-}
		Hydrogen sulfate ion,	HSO_4^-
		bisulfate	
		Sulfide	S^{2-}
		Hydrogen sulfide ion,	HS^-
		bisulfide	
		Sulfite	SO_3^{2-}
		Hydrogen sulfite ion,	HSO_3^-
		bisulfite	

[a] From *Chemistry: An Experimental Science*, edited by G. C. Pimentel. Copyright © 1963 by W. H. Freeman & Co. Reprinted with permission.

CHAPTER 1

1.8. (a) $\dfrac{[NH_3(g)]^2}{[N_2(g)][H_2(g)]^3}; \dfrac{[N_2(g)][H_2(g)]^3}{[NH_3(g)]^2}$

(b) $\dfrac{[NO(g)]^2}{[N_2O(g)][O_2(g)]^{1/2}}; \dfrac{[N_2O(g)][O_2(g)]^{1/2}}{[NO(g)]^2}$

(c) $\dfrac{[N_2(g)]^2[H_2O(g)]^6}{[NH_3(g)]^4[O_2(g)]^3}; \dfrac{[NH_3(g)]^4[O_2(g)]^3}{[N_2(g)]^2[H_2O(g)]^6}$

(d) $[NH_3(g)][HCl(g)]; \dfrac{1}{[NH_3(g)][HCl(g)]}$

1.9. 0.09 M.

1.10. 0.33 atm.

1.11. 278.

1.12. 4.8×10^{-3}. *Hint:* Do not forget to express the reactant and product as *concentrations* before substituting into Eq. (1.6).

1.13. 7.66 g of $N_2O_4(g)$ and 1.49 g of $NO_2(g)$. *Hint:* Proceed as in Exercise 1.6, and solve resulting quadratic equation.

1.14. These theorems follow in a fairly straightforward manner from the definition of the equilibrium constant [see Eq. (1.6)].

1.15. 3.6×10^{-3} and 17. *Hint:* Use the theorems stated in Exercise 1.14.

1.16. 0.14. *Hint:* Use one of the theorems stated in Exercise 1.14.

1.17. 2×10^{-4} atm each. *Hint:* The partial pressure of a solid can be equated to unity in Eq. (1.9b).

1.18. 2.1×10^{-5} atm; $CO_2(g)$. *Hint:* The partial pressure exerted by a gas in a mixture of gases is equal to the fractional contribution that the gas makes to the total volume of the mixture multiplied by the total pressure exerted by the mixture.

1.19.   *Solution:*

$$H_2O(l) \rightleftharpoons H_2O(g)$$

Since we can write $[H_2O(l)] = 1$ in Eq. (1.6) and $p_{H_2O(l)} = 1$ in Eq. (1.9b)

$$K_c = [H_2O(g)]$$

and,

$$K_p = p$$

where p is the partial pressure exerted by the water vapor in air.

From Eq. (1.9c)

$$\Delta n = 0 - 1 = -1$$

Therefore, from Eq. (1.9a)

$$p = [H_2O(g)]R_c^*T.$$

1.20. $H_2O(g)$ would decrease. *Hint:* Apply LeChatelier's principle.

1.21. $\dfrac{4f^2p}{1-f^2}$. *Hint:* The partial pressure exerted by a gas in a mixture of gases is equal to the fractional contribution that the gas makes to the total number of moles in the mixture multiplied by the total pressure exerted by the mixture.

1.22.   0.693 atm.
         Solution:

$$NH_4HS(s) \rightleftharpoons NH_3(g) + H_2S(g) \qquad \text{(A)}$$

From Eq. (1.9b), remembering that the partial pressures of solids can be equated to zero in this equation,

$$K_p = 1.08 \times 10^{-1} = p_1 p_2 \qquad \text{(B)}$$

where p_1 and p_2 are the partial pressures of $NH_3(g)$ and $H_2S(g)$, respectively, that are in equilibrium with $NH_4HS(s)$. The total pressure p produced by these gases is

$$p = p_1 + p_2 \qquad \text{(C)}$$

For every mole of $NH_3(g)$ released by reaction (A), a mole of $H_2S(g)$ is released. Therefore, the amounts of the two gases released by reaction (A) exert the same pressure. In the case of $NH_3(g)$, however, there is an additional pressure p_i' due to the 0.3 g of this gas that are already present in the flask. Therefore, we can write

$$p_1 = p_2 + p_i' \qquad \text{(D)}$$

We can calculate p_i' from the gas equation in the form

$$p_i'v = m\frac{R^*}{M}T$$

where v is the volume occupied by mass m of a gas with molecular weight M at temperature T. For the $NH_3(g)$ already present in the flask, $v = 2.00 \times 10^{-3}$ m^3, $m = 0.300 \times 10^{-3}$ kg, $M = 17$, $T = 298$K and $R^* = 8.31$ J deg^{-1} mol^{-1}. Hence, $p_i' = 0.219 \times 10^5$ Pa $= 0.219$ atm.

From Reactions (B) and (D), with $p_i' = 0.219$ atm we get

$$(p_2 + 0.219)p_2 = 1.08 \times 10^{-1}$$

or,

$$(p_2)^2 + 0.219p_2 - 0.108 = 0$$

Solution of this quadratic equation yields $p_2 = 0.237$ atm. From Reactions (C) and (D), with $p_i' = 0.219$ atm and $p_2 = 0.237$ atm, we get

$$p = 2p_2 + p_i' = 0.693 \text{ atm}$$

Therefore, the total pressure in the flask after chemical equilibrium is established is 0.693 atm.

1.23.   Since the reaction is endothermic it "moves to the right" with increasing temperature.

1.24.   At lower temperatures the reaction "shifts back to the left," but the rate of the reaction is now slow.

CHAPTER 2

2.11.   *Hint:* Remember that $\ln x = 2.303 \log_{10}x$.

2.12. 2.57×10^{-6}. *Hint:* Use expression given in Exercise 2.11.

2.13.   (a) 181 kJ; endothermic. (b) -199 kJ; exothermic. (c)

-196.2 kJ; exothermic. *Hint:* The enthalpies of formation of elements in their standard states are zero.

2.14. -74.8 kJ mol^{-1}. *Hint:* If a set of chemical reactions can be algebraically combined to yield a net reaction, the standard molar enthalpy of the net reaction can be obtained by similarly combining the standard molar enthalpies of the constituent reactions.

2.15.   *Hint:* Show that if a small quantity of heat does flow *from* a cold body *to* a hot body the total entropy of the system is *decreased*. Since this violates the second law of thermodynamics [see Eq. (2.20)], heat cannot flow from a cold to a hot body in an isolated system.

2.16.   In all three cases the entropy decreases.

2.17.   (a) Spontaneous at all temperatures. (b) Not spontaneous at any temperature. (c) Spontaneous at temperatures below a certain value. (d) Spontaneous at temperatures above a certain value.

2.18.   *Hint:* In addition to the two relations given, you will need to use Eqs. (2.6), (2.16), and (2.27).

2.19.   *Hint:* Follow analogous steps to those leading from Eq. (2.25) to Eq. (2.30).

2.21. $\Delta \overline{G}_f^0 = -69.6$ kJ; $K_p = 1.67 \times 10^{12}$; products are favored.

2.22.   *Solution:* From values of $\overline{S}^0$ and $\Delta \overline{H}_f^0$ given in Appendix V, it is found that $\Delta \overline{S}^0$ and $\Delta \overline{H}_{rx}^0$ are both negative for the stated reaction. Therefore, from the Gibbs–Helmholtz equation (2.29), the forward reaction is more likely to produce a decrease in the Gibbs free energy, which favors dissociation of NO(g), at low temperatures. At the high temperatures present in combustion, the reverse reaction, which forms the pollutant NO(g), will be favored.

2.24. $K_p(298K) = 3.92 \times 10^{21}$; $K_p(500K) = 1.27 \times 10^{14}$. *Hint:* Use the relation given in Exercise 2.11 to find K_p at 500K.

2.25.   Decrease of 3×10^{-21} J.

2.27.   *Hint:* Start with Eq. (2.8). For solids and liquids (but not gases) dV and dp are small near 1 atm.

2.28.   *Hint:* Combine the gas equation for a unit mass [see Eq. (1.8c)] with Eq. (2.6) and the definition of c_p.

2.29. $\Delta G = -nF \Delta E$

Solution: From Exercise 2.18 we have

$$da = -dg$$

where da is the work done over and above any $pd\alpha$ work. In this case, this extra work is electrical. Therefore,

$$\Delta G = - \text{(electrical work done)}$$
$$= - \text{(voltage difference) (charge transfer in coulombs)}$$
$$= - \Delta E \, (nF)$$

or,

$$\Delta G = - \, nF \, \Delta E$$

CHAPTER 3

3.8. $m=3$, $n=1$, $k=8.3\times10^{-4}$ M^{-3} s^{-1}; reaction rate $=$ 6.6×10^{-7} M s^{-1}.

3.10.   (a) $k_{\text{pseudo}}=1.8\times10^{-20}$ [O$_3$(g)]. (b) $k_{\text{pseudo}}=7.3\times10^{-3}$ s^{-1}.

3.11.   *Hint:* Combine Eqs. (3.5) and (1.8d).

3.12.   (a) $2O_3(g)\rightarrow3O_2(g)$. (b) O(g). (c) For step (i): Rate $=k_1$ [O$_3$(g)]. For step (ii): Rate $=k_2$ [O$_3$(g)]‖O(g)]. (d) Step (ii). (e) [O(g)] $\propto$ [O$_3$(g)][O$_2$(g)]$^{-1}$.

3.13.   *Hint:* In each case, write an expression for $-\dfrac{d[O_2(g)]}{dt}$ for the rate-determining step; then apply Eq. (1.6) to the fast reaction.

3.14. $K_{\text{c}}=\dfrac{k_{\text{f}}}{k_{\text{r}}}=\dfrac{1.5\times10^{-12}}{5.0\times10^{-2}}=3.0\times10^{-11}$

3.15.   (a) Reaction mechanism (iii).

     (b) $k=\dfrac{k_{4\text{f}}k_{5\text{f}}}{k_{4\text{r}}}$

     (c) $K_{\text{c}}=\dfrac{k_{4\text{f}}k_{5\text{f}}k_{6\text{f}}}{k_{4\text{r}}k_{5\text{r}}k_{6\text{r}}}$

     Hint: Apply the principle of detailed balancing. See, for example, Exercise 3.4.

3.16. 4.8×10^7 mol L^{-1} s^{-1}; 5.0×10^{-10} s.

3.18. $E_{\text{a}}=53.3$ kJ mol^{-1}.

3.19.   Rate increases by a factor of 1.8.

3.20. $E_{\text{a}}=115$ kJ mol^{-1}; $A=10^{25}$ s^{-1}.

3.21.   1.25 g. *Hint:* The amount of the reactant will halve at the end of each period of time equal to one half-life.

3.23. 1.9×10^3 years. *Solution:* Since radioactivity is a first-order chemical reaction, we have

$$\text{Decay rate} = \lambda N \qquad \text{(A)}$$

where λ = decay constants [equivalent to k in Eq. (3.4)]. Hence, by comparison with Eqs. (3.4), (3.5), and (3.10)

$$\log[N]_t = \frac{-\lambda t}{2.303} + \log[N]_0 \qquad \text{(B)}$$

and

$$\lambda = \frac{0.693}{t_{1/2}} \qquad \text{(C)}$$

For carbon-14, $t_{1/2} = 5.7 \times 10^3$ yr; therefore, from (C)

$$\lambda = \frac{0.693}{5.7 \times 10^3} = 1.2 \times 10^{-4} \text{ yr}^{-1} \qquad \text{(D)}$$

From (A)

$$[N] = \frac{\text{Decay rate when carbon-14 equilibrium ceased}}{\lambda} = \frac{15}{\lambda} \qquad \text{(E)}$$

$$[N]_t = \frac{\text{Decay rate at present time } (t)}{\lambda} = \frac{12}{\lambda} \qquad \text{(F)}$$

From (B)

$$\log\frac{[N]_t}{[N]_0} = -\frac{\lambda t}{2.303}$$

and using (D) – (F)

$$\log\frac{12}{15} = \frac{-(1.2 \times 10^{-4})t}{2.303}$$

Therefore,

$$t = 1.9 \times 10^3 \text{ yr}$$

3.24. $\tau(NH_3) = 19$ days; $\tau(N_2O) = 15$ yr; $\tau(CH_4) = 12$ yr.

3.25.   (a) 1 hr. (b) 20 hr. *Hint:* Use Eq. (3.12); read first complete sentence in text that follows Eq. (3.12).

CHAPTER 4

4.10.   2.1%; 0.21 M; 0.24 m; 0.019

4.11.   *Hint:* Apply LeChatelier's principle.

4.12.   1.56 m. *Hint:* Apply Henry's law.

4.13.   2.37 kg. *Hint:* Apply Raoult's law.

4.14.   Three parts of glycol to 10 parts of water. *Hint:* Apply Eq. (4.7).

4.15. $Ba(OH)_2$, KNO_3, H_2SO_4, and HCl are soluble in water; the remaining have low solubilities. *Hint:* See Table 4.1.

4.16.   660 kJ.

4.17.   (a) $[Ag^+(aq)]^2[CrO_4^{2-}(aq)]$. (b) $[Ca^{2+}(aq)][SO_4^{2-}(aq)]$. (c) $[H^+(aq)][CH_3COO^-(aq)]$.

4.18. 5.3×10^{-3} g L^{-1}.

Solution:

$$CaCO_3(s) \rightleftarrows Ca^{2+}(aq) + CO_3^{2-}(aq)$$

Therefore, for each mole of $CaCO_3(s)$ that dissolves in water, 1 mole of $Ca^{2+}(aq)$ and 1 mole of CO_3^{2-} (aq) enter the solution. Let x be the solubility of $CaCO_3(s)$ in moles per liter, then the molar concentrations of $Ca^{2+}(aq)$ and CO_3^{2-} (aq) will each be x. Therefore, since,

$$K_{sp} = [Ca^{2+}(aq)][CO_3^{2-}(aq)] = 2.8 \times 10^{-9}$$

$$[x]^2 = 2.8 \times 10^{-9}$$

or,

$$[x] = 5.29 \times 10^{-5}$$

Hence, the solubility of $CaCO_3(s)$ is 5.29×10^{-5} M. Therefore, 5.29×10^{-5} moles of $CaCO_3(s)$ dissolves in 1 L of water. Or, 1 mole of $CaCO_3(s)$ dissolves in $1/(5.29 \times 10^{-5})$ or 1.89×10^4 L of water. Since the molecular weight of $CaCO_3$ is 100.1, 100.1 g of $CaCO_3(s)$ dissolve in 1.89×10^4 L of water. Therefore, $\dfrac{100.1}{1.89 \times 10^4}$ or 5.29×10^{-3} g of $CaCO_3(s)$ dissolves in 1 L of water, which is therefore the solubility of $CaCO_3(s)$ in water.

4.19. 4.3×10^{-13} M.

4.20.   (a) 1.3×10^{-4} M. (b) 1.6×10^{-7} M. *Hint:* Lead nitrate acts as a source of $Pb^{2+}(aq)$ ions in the solution, as does $PbSO_4(s)$.

Therefore, this is a "common ion" problem; proceed as in Exercises 4.4 and 4.5.

CHAPTER 5

5.8.   Concentrations of $H^+(aq)$ and $OH^-(aq)$ are 2×10^{-13} and 0.05 M, respectively. NaOH is a base.

5.9.   Concentrations of $H^+(aq)$ and $OH^-(aq)$ are 0.040 and 2.5×10^{-13} M, respectively.

5.10.   (a) $HSO_4^-(aq) + H_2O(l) \rightleftarrows H_3O^+(aq) + SO_4^{2-}(aq)$
   acid 1 base 2 acid 2 base 1

  (b) $H_2PO_4^-(aq) + HCl(l) \rightleftarrows H_3PO_4(aq) + Cl^-(aq)$
   base 2 acid 1 acid 2 base 1

  (c) $NH_4^+(aq) + CH_3COO^-(aq) \rightleftarrows CH_3COOH(aq) + NH_3(aq)$
   acid 1 base 2   acid 2   base 1

5.11.   0.010 and 1.0×10^{-12} M, respectively. pH $= 12$.

5.12.   Acid-dissociation constant is 3.96×10^{-10}.

Solution: Let y be the number of moles per liter of HCN that ionize.

$$HCN(g) + H_2O(l) \rightleftarrows H_3O^+(aq) + CN^-(aq)$$

Initially in solution: 0.200 M
Changes: $-y$M yM yM
At equilibrium: $(0.200 - y)$M yM yM

Also, $y = [H_3O^+(aq)] = [CN^-(aq)]$.
Therefore,

$$-\log[H_3O^+(aq)] = -\log y = pH = 5.05$$

or,

$$y = 8.91 \times 10^{-6}$$

Hence,

$$\text{Acid-dissociation constant for HCN} = \frac{[H_3O^+(aq)][CN^-(aq)]}{[HCN(g)]}$$

$$= \frac{y^2}{0.2 - y}$$

$$= \frac{(8.9 \times 10^{-6})^2}{0.2 - 8.9 \times 10^{-6}}$$

$$= 3.96 \times 10^{-10}$$

5.13. Concentration of protons is 1.3×10^{-3} M. *Hint:* Proceed in similar manner to Exercise 5.12.

5.14. $[H^+(aq)] = [H_2PO_4^-(aq)] = 8.3 \times 10^{-3}$ M; $[HPO_4^{2-}(aq)] = 6.2 \times 10^{-8}$ M; $[PO_4^{3-}(aq)] = 3.6 \times 10^{-18}$ M

Hint: Assume that the $H^+(aq)$ ions derive mainly from the first stage of dissociation, and that the concentration of any ion formed in one stage is not significantly affected by succeeding dissociations.

5.15. pH = 7; fraction hydrolyzed = 1.1%. *Solution:* Since $K_a(HC_2H_3O_2) = K_b(NH_3)$, the cations and anions hydrolyze equally. Therefore, the concentration of $H_3O^+(aq)$ due to hydrolysis of $NH_4^+(aq)$ will be the same as the concentrations of OH^- ions from the hydrolysis of $C_2H_3O_2^-$. Therefore, the pH of the solution is 7, and $[H_3O^+(aq)] = 10^{-7}$.

For the $NH_4^+(aq)$ hydrolysis,

$$K_h = K_w/K_b(NH_3)$$
$$= 10^{-14}/1.8 \times 10^{-5} = 5.6 \times 10^{-10}$$

The hydrolysis reaction for NH_4^+ is

$$NH_4^+(aq) + H_2O(l) \rightleftharpoons NH_3(aq) + H_3O^+(aq)$$

Therefore,

$$K_h = \frac{[NH_3(aq)][H_3O^+(aq)]}{[NH_4^+(aq)]} = 5.6 \times 10^{-10}$$

Let, $x = [NH_3(aq)]$. Then,

$$0.0050 - x = [NH_4^+(aq)]$$

Therefore,

$$\frac{10^{-7}x}{0.0050 - x} = 5.6 \times 10^{-10}$$

Let us assume that $0.0050 - x \simeq 0.0050$, then

$$\frac{10^{-7}x}{0.0050} = 5.6 \times 10^{-10}$$

or, $x = 2.8 \times 10^{-5}$ (which justifies our assumption that $x \ll 0.0050$). Hence, the quantity of $NH_4^+(aq)$ that is hydrolyzed is 2.8×10^{-5} M. Since $K_a(HC_2H_3O_2) = K_b(NH_3)$,

$K_h(= K_w/K_a)$ for the hydrolysis of $C_2H_3O_2^-(aq)$ has the same value as that calculated above for NH_3, namely, $5.6 \times 10^{-10.}$ Therefore, the amount of $C_2H_3O_2^-(aq)$ hydrolyzed is also 2.8×10^{-5} M. Therefore,

Fraction of $NH_4C_2H_3O_2$ hydrolyzed

$$= \frac{\text{Amount of } NH_4^+(aq) \text{ hydrolyzed} + \text{amount of } C_2H_3O_2^-(aq) \text{ hydrolyzed}}{\text{Original amount of } NH_4C_2H_3O_2}$$

$$= \frac{2(2.8 \times 10^{-5})}{0.0050}$$

$$= 1.1 \times 10^{-2} \text{ or } 1.1\%$$

5.16.   pH $= 1.1$. *Hint:* Every mole of $Na^+(aq)$ releases 1 mole of negative ions [e.g., $Cl^-(aq)$ or $OH^-(aq)$]. Similarly, every mole of $Ca^{2+}(aq)$ releases two moles of negative ions, and every mole of $Cl^-(aq)$ releases one mole of positive ions [e.g., $Na^+(aq)$ or $H^+(aq)$], and every mole of $SO_4^{2-}(aq)$ releases two moles of positive ions. To solve the problem, find the total number of moles of negative charge released and the total number of moles of positive charge released, and hence the net number of moles of charge released. Since the only other charge carriers are $H^+(aq)$ and $OH^-(aq)$, the net number of moles of positive or negative charge must be in the form of $H^+(aq)$ or $OH^-(aq)$ ions; this will permit determination of the pH of the solution.

5.17.   pH $= 8.95$. *Solution:* NH_4^+ (from NH_4Cl) and NH_3 are an acid-base conjugate pair. From Eq. (5.26)

$$\frac{[NH_4^+]}{[NH_3]} = \frac{[H_3O^+(aq)]}{K_a(NH_4^+)}$$

Therefore,

$$[H_3O^+(aq)] = K_a(NH_4^+)\frac{[NH_4^+]}{[NH_3]}$$

Since $K_a(NH_4^+) = 5.60 \times 10^{-10}$ and,

$$\frac{[NH_4^+]}{[NH_3]} = 2.00$$

$$[H_3O^+(aq)] - (5.60 \times 10^{-10}) \times (2.00)$$

$$= 11.2 \times 10^{-10}$$

$$\therefore pH = -\log[H_3O^+(aq)]$$

$$= -\log(11.2 \times 10^{-10})$$

$$= -(1.05 - 10)$$

Therefore,

$$pH = 8.95$$

5.18.   (a) 2.6×10^{-4} M. (b) pH would decrease by 0.22 pH units. (c) pH would increase by 0.14 pH units. (c) 2.6×10^{-4} M. *Hints:* See Exercises 5.4 and 5.5.

5.19.   Range of pH values is 3.7 to 6.5. A pH of 6.5 would not be achieved because buffering capacity of CO_2 in the air would hold pH to ≤ 5.7.

CHAPTER 6

6.12.   (a) The oxidation half-reaction is

$$Cu^+(aq) \rightarrow Cu^{2+}(aq) + e^-$$

The reduction half-reaction is

$$Fe^{3+}(aq) + e^- \rightarrow Fe^{2+}(aq)$$

$Fe^{3+}(aq)$ is the oxidant and $Cu^+(aq)$ the reductant. (b) The oxidation half-reaction is

$$Zn(s) \rightarrow Zn^{2+}(aq) + 2e^-$$

The reduction half-reaction is

$$2H^+(aq) + 2e^- \rightarrow H_2(g)$$

$H^+(aq)$ is the oxidant and $Zn(s)$ the reductant.

6.13.   (a) Oxidation number of oxygen and nitrogen are -2 and 3, respectively. (b) Oxidation number of oxygen and nitrogen are -2 and 5, respectively. (c) Oxidation number of hydrogen, oxygen and nitrogen are 1, -2, and 5, respectively. (d) Oxidation number of hydrogen, oxygen, and sulfur are 1, -2, and 4, respectively.

6.14.   (a) Yes. (b) No. *Hint:* Determine whether or not any of the oxidation numbers of any of the atoms is changed by the reaction.

6.15. $2Pb(s) + O_2(aq) + 2H_2O(l) \rightarrow 2Pb^{2+}(aq) + 4OH^-(aq)$

6.16. $4P_4(s) + 12OH^-(aq) + 12H_2O(l) \rightarrow 12H_2PO_2^-(aq) + 4PH_3(aq)$.
Hint: See Exercises 6.2–6.4.

6.17. $O_2(g)$ is the oxidant and $C_6H_{12}O_6(s)$ the reductant.

6.18. $Mn^{2+}(aq)$, $Fe^{2+}(aq)$, $Ag^+(aq)$, $O_3(aq)$. *Hint:* See Table 6.2 and associated discussion in the text.

6.19.   (a) Yes, it will. (b) A reductant. (c) Both. (d) No, it will not. *Hint:* See Table 6.2 and associated discussion in the text.

6.20.   (a) It will. $E^0_{cell} = 1.22$ V. (b) It will not. $E^0_{cell} = -0.781$ V. *Hint:* See Exercise 6.6.

6.21. $\Delta \overline{G}^0 = 1.05 \times 10^5$ J; it is not a spontaneous reaction. *Hint:* Use Eq. (6.20).

6.22. $K_c = 4.57 \times 10^{-19}$. The equilibrium state for the reaction lies far to the left.

6.23. $E_{cell} = 0.769$ V; $\Delta G = -1.48 \times 10^5$ J; the forward reaction is spontaneous. *Hint:* Use the Nernst equation (6.25) to obtain E_{cell} at 250K from E^0_{cell}. Use an analogous expression to Eq. (6.20) to obtain ΔG at 250K.

6.24.   Maximum voltage = 12.246 V.

6.25. $[H^+(aq)] = 1.3 \times 10^{-3}$ M. *Hint:* Use Eq. (6.26).

6.26. $E^0_{cell} = -0.046$ V. The reaction is not spontaneous under standard conditions.
Solution: $HSO_3^-(aq) + H_2O(l) \rightarrow$
$\qquad\qquad HSO_4^-(aq) + 2H^+(aq) + 2e^-$
From Eq. (6.20)

$$\Delta \overline{G}^0 = -nFE^0_{cell}$$

Therefore,

$$E^0_{cell} = -\frac{\Delta \overline{G}^0}{nF}$$

where $n = 2$ and $F = 96,489$ C and, from Eq. (2.34), $\Delta \overline{G}^0 = -227$ kJ mol^{-1}. Hence, $E^0_{cell} = 1.2$ V. Since $\Delta \overline{G}^0$ is negative (and E^0_{cell} positive), the reaction is spontaneous under standard conditions.

6.27. $\dfrac{Fe^{2+}(aq)}{Fe^{3+}(aq)} = 782$

Solution: From Table 6.2 we see that the standard potential for the $Fe^{3+}(aq)$ – $Fe^{2+}(aq)$ couple is 0.771 V. Since the seawater has a redox potential of 0.600 V, it provides a better reducing environment than the $Fe^{3+}(aq)$ – $Fe^{2+}(aq)$ couple. Therefore, this couple will be reduced; that is, it will be driven from left to right.

$$Fe^{3+}(aq) + e^- \rightarrow Fe^{2+}(aq) \qquad E_{red}^0 = 0.771 \text{ V}$$

The seawater will be involved in the *oxidation half-cell reduction* (i.e., it acts as the reductant) with $E_{ox} = -(0.600 \text{ V}) = -0.600 \text{ V}$. When the seawater is in equilibrium with the iron system $E_{cell} = E_{red} + E_{ox} = 0$ or $E_{red} = 0.600 \text{ V}$. For nonstandard concentrations at 298K, the total cell potential developed by the iron system when paired with the hydrogen half-cell is given by Eq. (6.26). However, the hydrogen half-cell generates zero electrode potential. Therefore, the electrode potential developed by the iron system is, from Eq. (6.26)

$$E_{red} = E_{red}^0 - \frac{0.0591}{1} \log \frac{[Fe^{2+}(aq)]}{[Fe^{3+}(aq)]}$$

or,

$$0.600 = 0.771 - 0.0591 \log \frac{[Fe^{2+}(aq)]}{[Fe^{3+}(aq)]}$$

Therefore,

$$\frac{Fe^{2+}(aq)}{Fe^{3+}(aq)} = 782$$

6.28.   1 equiv. of $H_2S(aq)$ is 17 g, and 1 equiv. of $HNO_3(aq)$ is 21 g. *Hint:* From Exercise 6.11 we have

$$\text{Oxidation number of a species} = \frac{1 \text{ mole of the species}}{1 \text{ equiv. of the species}}$$

6.29.   Normality is 0.57 N.

CHAPTER 7

7.7.   Ratios of energies are 1: 0.08: 0.8: 0.004.

7.8. 4×10^5 J.

7.9.   For nitrogen, $\lambda_{max} = 0.08$ μm; for oxygen, $\lambda_{max} = 0.099$ μm. Both these wavelengths lie in the ultraviolet region.

7.10. 2.9×10^7 photons.

7.11.   34%.

7.12. $$\frac{[NO_2(g)]}{[NO(g)]} = 2.5$$

Solution: From Eq. (7.23)

$$\frac{[NO_2(g)]}{[NO(g)]} = \frac{1.0 \times 10^{-20}}{4.0 \times 10^{-3}} [O_3(g)]$$

where $[O_3(g)]$ must be in units of molecules m^{-3}. The concentration of O_3 is given in ppmv; to change this to molecules m^{-3}, we must first apply the gas equation in the form of Eq. (1.8g) to air at a pressure of 1 atm (1013×10^2 Pa) and a temperature of 20°C or 293K, which gives $n_0 = 2.5 \times 10^{25}$ molecules m^{-3}, where n_0 is the concentration of all the molecules in air at 1 atm and 20°C. Since $O_3(g)$ occupies 0.040 ppmv of air, $[O_3(g)] = (2.5 \times 10^{25})(0.040 \times 10^{-6}) = 1.0 \times 10^{18}$ molecules m^{-3}. Hence,

$$\frac{[NO_2(g)]}{[NO(g)]} = \frac{1.0 \times 10^{-20}}{4.0 \times 10^{-3}} (1.0 \times 10^{18}) = 2.5$$

7.13. $$\frac{dn_1}{dt} = 2j_a n_2 - k_b n_1 n_2 n_M + j_c n_3 - k_d n_1 n_3$$

$$\frac{dn_2}{dt} = -j_a n_2 - k_b n_1 n_2 n_M + j_c n_3 + 2k_d n_1 n_3$$

$$\frac{dn_3}{dt} = k_b n_1 n_2 n_M - j_c n_3 - k_d n_1 n_3$$

7.14.   Result could have been predicted because $n_1 + 2n_2 + 3n_3$ is the sum of the oxygen atoms, which cannot change. *Hint:* To obtain $n_1 + 2n_2 + 3n_3 =$ constant, add the left and right sides of the three differential equations given above for the solution of Exercise 7.13, and then integrate.

7.15.   When $X = H$

$$H + O_3 \rightarrow OH + O_2$$
$$OH + O \rightarrow H + O_2$$

Net: $O + O_3 \rightarrow O_2 + O$

When $X = OH$

$$OH + O_3 \rightarrow HO_2 + O_2$$
$$HO_2 + O \rightarrow OH + O_2$$

Net: $\quad O + O_3 \rightarrow O_2 + O_2$

7.16. The rate coefficients for Reactions (7.27) and (7.36b) can never be the same. The rate coefficient for Reaction (7.36b) is the larger.

Index

189